COPYRIGHT 2020

ACKNOWLEDGEMENTS

The author wishes to thank Sally J. Bensusen, Sr. Graphic Designer NASA Science Program Support Office, Global Science and Technology, Inc. for providing me with links to the satellite images used in this book. These images *are not* photographs, but images created from data acquired by a satellite and made into "visualizations" by the various imaging teams. All images have been used via the courtesy of NASA".

In addition, Simon Balm's computer assistance was indispensable. His help in dealing with Kindle Publishing made this book possible. Also, Brian Pena has been a lifesaver solving many problems on many occasions! This book could not have been made without their assistance.

FORWARD

- This book, written for the *general public*, will give you mile by mile descriptions of the geology from Kahului to Haleakala's summit and from Hana to the Kipahulu Area of Haleakala National Park.
- In some areas, the geology is very simple, and a brief description will be given at the indicated mileage points.
- In other areas, the geology is more complex, and the main highlights will be given, followed by, in a few cases, optional details.

ODOMETERS

- Try to set your odometer as close as possible to the starting point given in the road logs.
- However, all odometers are not created equal. Odometer deviations of 0.1-0.3 miles are not uncommon.
- However, these minor differences should not cause you any serious problems.

MILEAGE DEVIATIONS

- Any deviations from the road log mileages must be added to those given in the road log.

DISCLAIMER

- **Use caution and common sense while using these geologic road logs.**
- **Drive, park, and hike safely.**
- **Do not go any place you may feel is unsafe.**
- **Do not venture away from the beaten path.**
- **After you park your car, do not leave valuables in plain sight, even if the car is locked.**
- **The author and publisher assume no liability for accidents, injury, or any losses or deaths by individuals or groups using this publication.**
- **The author of this geology guide assumes no responsibility for any accidents, injuries, or deaths based on these descriptions.**
- *Users of this book must assume full responsibility for their actions.*

ABOUT THE AUTHOR

- The author of this book taught geology at a southern California college for 41 years.
- Many trips were taken to Maui to make these road log mileage points as accurate as possible.

ADDITIONAL BOOKS BY THIS AUTHOR

- Geological Guide to Oahu (color version)
- Geology of Oahu (black and white version)
- Illustrated Guide to the Island of Hawaii (color version)
- Island of Hawaii Geological Guide (Black and white version)
- Geological Guide to Hawaii Volcanoes National Park (6x9 color version)
- A Guide to the Geology and Geography of Kauai (color version)
- Geology and Geology of Kauai (black and white version)
- *All of the above books are available on Amazon.*
- *All prices set by Amazon.*

IMPORTANT TRAVEL INFORMATION

- Before beginning these road logs, be sure to *start with a full tank of gas!*
- Also, even if you have a GPS unit, be sure to also have a *good* road map. GPS units are *not* infallible!
- In the event of an emergency, be sure your cell phone is charged.
- However, in some rural areas you may not have cell phone service.
- Loss of cell phone service could be especially true on the south side of Haleakala.
- Also, have your roadside assistance phone numbers available should the need arise.
- These phone numbers are generally on your rental car's key ring.
- Above all, watch for careless drivers and be sure that you are not one yourself.
- At times, to prevent a particular mileage point from catching you "off guard", some of the mileages to a particular location have been shorten by 0.1 miles.

HAWAIIAN ISLAND DIRECTIONS

- Note: you will hear locals using *mauka* and *makai* when giving directions.
- Islanders do not use the terms inland and seaward.
- *Mauka* is used for *inland*.
- *Makai* is used for *seaward*.

The following websites will allow you to speak like an islander

To *hear* Maui *place names* pronounced, go to:

http://hawaiian-words.com/hawaii-place-names

To *hear* Hawaiian *words* pronounced, go to:

http://hawaiian-words.com/common

....

TABLE OF CONTENTS

MAP OF MAUI COUNTY: p. 1

MAP OF MAUI: p.2

PHYSIOGRAPHIC IMAGE OF MAUI: p 2

SATELLITE IMAGE OF MAUI: p. 3

LOCATION OF TOWNS ON MAUI: p. 3

ESTIMATED TRAVEL TIMES BETWEEN MAUI TOWNS: p. 4

IMPORTANT TRAVEL INFORMATION: p.5

WORDS USED FOR HAWAIIAN ISLAND DIRECTIONS: p. 5

WEBSITES TO HELP PRONOUNCE MAUI *PLACE NAMES*: p. 5

WEBSITES TO HEAR HAWAIIAN *WORDS* PRONOUNCED: p.5

MAUI AND HALEAKALA TIMELINE: p. 6

WEBSITE FOR MAUI NEWS: p. 8

BRIEF MAUI OVERVIEW: p. 9

GEOGRAPHY OF HAWAIIAN ISLANDS AND MAUI: p. 10

COMMUNITY DISTRICTS OF MAUI: p. 12

ELEVATION MAP OF MAUI: p. 13

CLIMATE OF MAUI: p. 14

DISTRIBUTION OF MAUI VEGETATION VERSUS RAINFALL: p. 14

GEOGRAPHIC DISTRIBUTION OF MAUI RAINFALL: p. 15

SUMMARY OF MAUI WEATHER: p. 15

DROWNINGS ON MAUI: p. 16

BEACH SAFETY WEBSITES: p. 16

WEBSITE OF SHARK ATTACKS ON MAUI: p. 16

SIMPLIFIED GEOLOGIC TIME SCALE: p. 17

ORIGIN OF THE HAWAIIAN ISLANDS: p. 18

GENERALIZED STAGES OF HAWAIIAN ISLAND VOLCANOES: p. 19

GEOLOGY OF MAUI: p. 20

VOLCANOES ON MAUI: p. 20

ORIGIN OF THE WEST MAUI VOLCANO: p. 21

GENERAL DESCRIPTION OF THE HALEAKALA VOLCANO: p. 21

RIFT ZONES: p. 22

CALDERAS: p. 22

VOLCANIC UNITS ON HALEAKALA: p. 23

GENERALIZED GEOLOGIC MAP OF MAUI: p. 24

HALEAKALA ERUPTION HAZARDS: p. 25

EARTHQUAKES ON MAUI: p. 26

TSUNAMI HAZARDS ON MAUI: p. 26

THE PLEISTOCENE GLACIATION ON MAUI: p. 27

PLEISTOCENE GLACIATION HALEAKALA: p. 28

OPTIONAL CHEMICAL COMPOSITION OF HAWAIIAN LAVA FLOWS: p. 28

OPTIONAL OCCURRENCE OF THE TYPES OF THESE TYPES OF LAVAS: 28

ELEVATION MAP OF MAUI: p. 28

GEOLOGY OF THE HALEAKALA VOLCANO: p. 29

HALEAKALA'S DEVELOPMENT: p. 30

HALEAKALA'S SUMMIT BASIN (aka, it's crater) p. 31

HALEAKALA'S ERUPTION HISTORY: p. 32

GEOLOGIC MAP OF HALEAKALA LAVA FLOWS: p. 32

MAP OF LAVA FLOWS IN THE SUMMIT BASIN: p. 33

HALEAKALA ERUPTION HAZARDS: p. 34

SUMMARY OF HALEAKALA WEATHER: p. 35

LOCATION MAP OF MAUI TOWNS IN RELATION TO THE RAINFALL: p. 35

HALEAKALA DURING THE PLEISTOCENE GLACIATION: p. 36

GEOLOGY OF EARLY HAWAIIAN SETTLEMENTS ON HALEAKALA: p. 36

GEOLOGY RELATED TO HAWAIIAN SETTLEMENTS: p. 37

GEOLOGY'S RELATIONSHIP TO THE EARLY HAWAIIAN SETTLEMENTS ON HALEAKALA IN THE KAHIKINUI REGION: p. 37

GEOLOGICAL GUIDE TO HALEAKALA NATIONAL PARK: p. 38

    PREPARING TO DRIVE TO HALEAKALA NATIONAL PARK: p. 39

    ROUTE FOR KAHULUI TO HALEAKALA NATIONAL PARK: p. 39

    VIEWING SUNRISE AND SUNSET ON HALEAKALA: p. 40

    SUMMIT PRECAUTIONS: p. 41

    GLOSSARY OF SELECTED VOLCANIC TERMS: p. 42

    THREE COMMON TYPES OF VOLCANOES: p. 45

    HAWAIIAN PRONUNCIATION GUIDE: p. 46

    MEANING AND PRONUNCIATION OF MAUI *PLACE NAMES:* p. 47

    HAWAIIAN WORDS YOU SHOULD KNOW: p. 48

    MISPRONOUNCED HAWAIIAN WORDS: p. 49

    WEBSITES TO PRONOUNCE MAUI *PLACE NAMES*: p. 49

    WEBSITES TO HEAR HAWAIIAN *WORDS* PRONOUNCED: P. 49

    WEBSITES FOR MAUI NEWS AND INFORMATION: p. 50

    TIPS FOR HIKING IN HAWAII: p. 50

    TYPES OF HAWAIIAN LAVA: p. 51

    STRATIGRAPHIC AGES OF VOLCANIC UNITS ON MAUI: p. 52

    SYMBOLS USED ON MAUI GEOLOGIC MAPS FOR SEDIMENTS: P.52

    SYMBOLS USED ON MAUI GEOLOGIC MAPS FOR HALEAKALA VOLCANICS: p. 52

    SYMBOLS USED ON MAUI GEOLOGIC MAPS FOR THE WEST MAUI VOLCANO: P. 52

    GEOLOGIC MAP OF SOUTHEASTERLY MAUI: p. 53

    GEOLOGIC MAP OF SOUTH CENTRAL MAUI: p. 54

    GEOLOGIC MAP OF CENTRAL SOUTHEASTERLY MAUI: p. 55

    GEOLOGIC MAP OF NORTH EAST MAUI: p. 56

    GEOLOGIC MAP OF NORTH EASTERLY MAUI: p. 57

    GEOLOGIC MAP OF EAST MAUI: p. 58

    GEOLOGIC MAP OF NORTH WESTERN MAUI: p. 59

    REFERENCES USED IN THIS BOOK: p. 60

MAP OF MAUI COUNTY

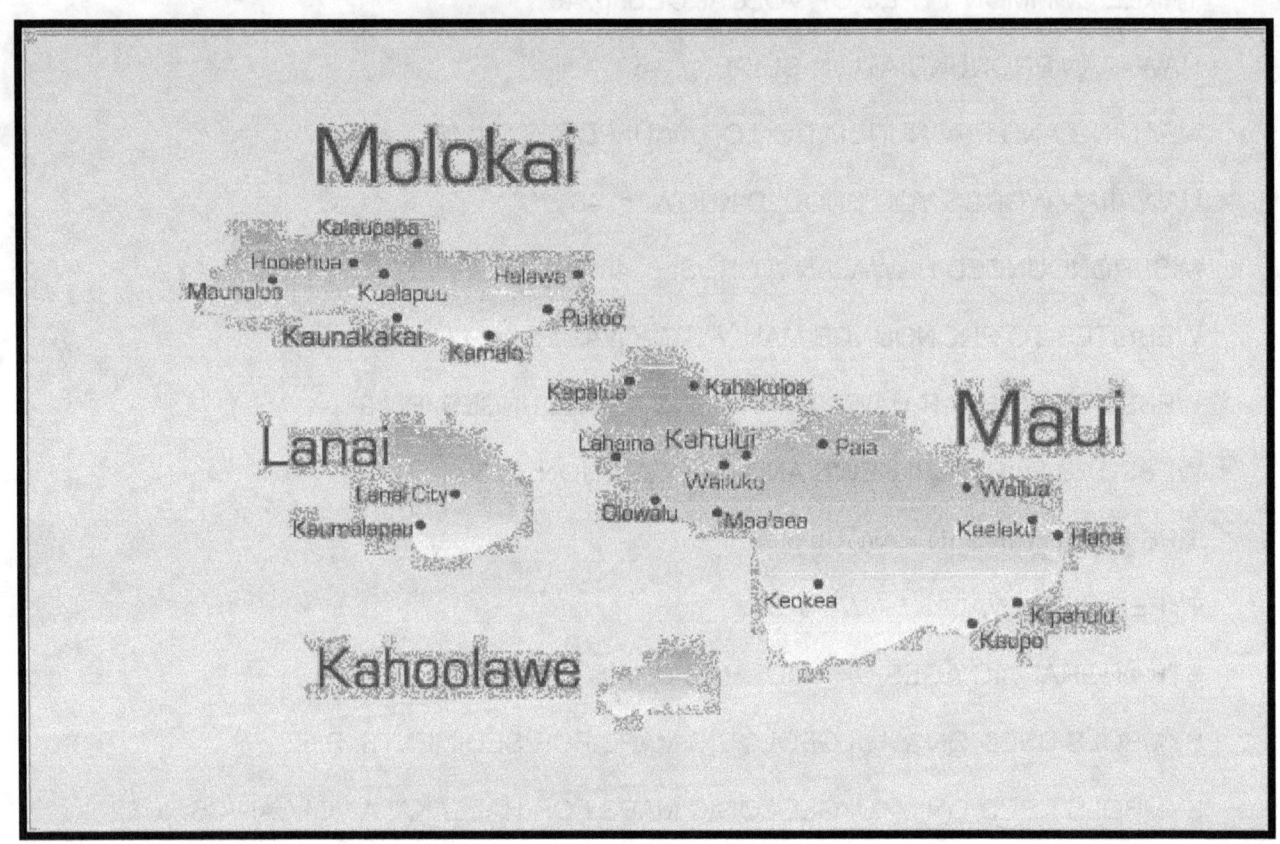

Courtesy of Maui County

Note: Maui County includes the islands of Maui, Molokai, Lanai, and Kahoolawe (uninhabited)

MAP OF MAUI

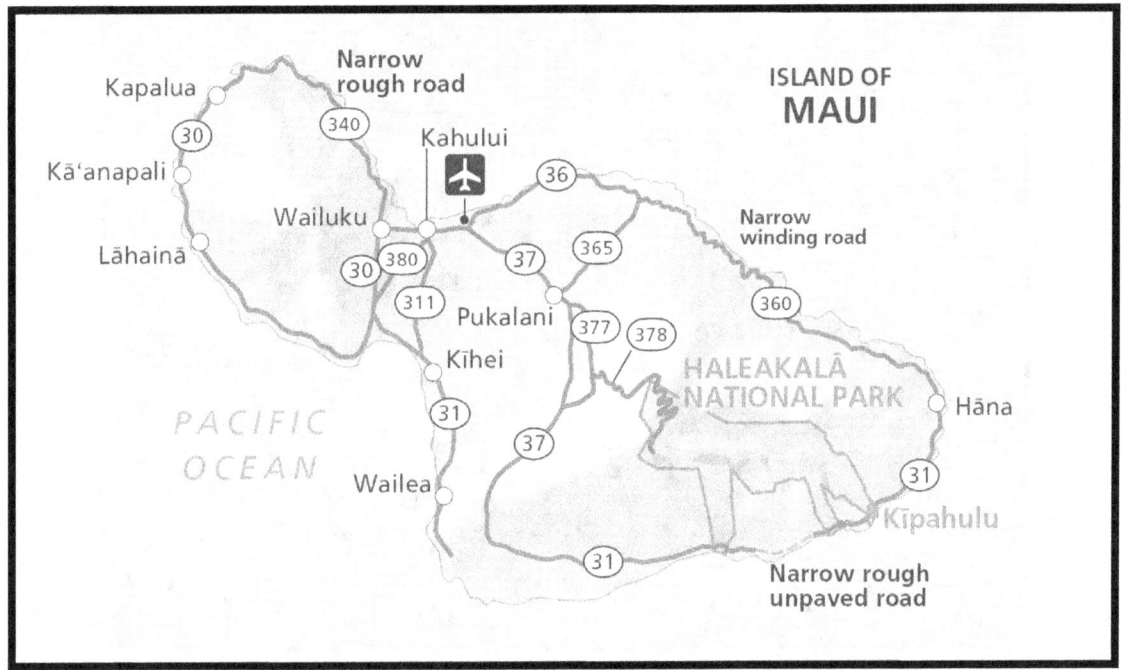

Courtesy of the National Park Service

PHYSIOGRAPHIC IMAGE OF MAUI

Courtesy of the National Park Service

SATELLITE IMAGE OF THE HAWAIIAN ISLANDS

COURTESY OF NASA

LOCATION OF TOWNS ON MAUI

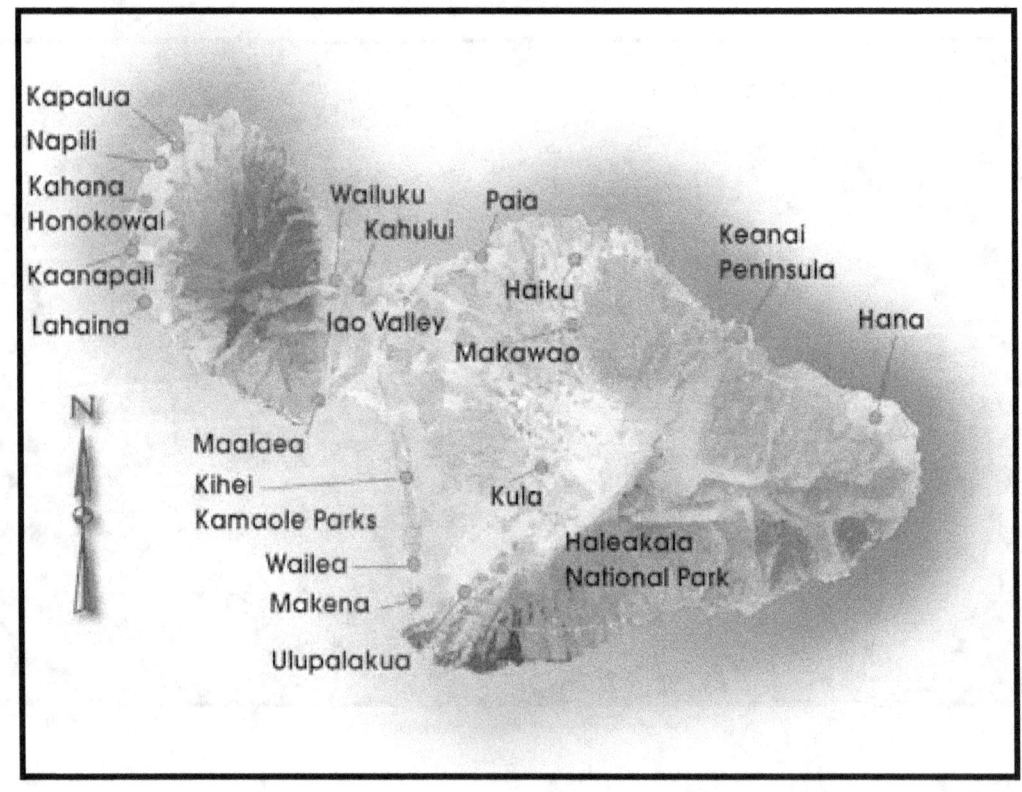

GENERALIZED MAP OF TOWNS AND ROADS ON MAUI

Modified from Maui Hotels.com

Estimated driving times on Maui (without traffic)

Kahului to Lahaina: 40 minutes
Kahului to Kaanapali: 45 minutes
Kahului to Kapalua: 55 minutes
Kahului to Maalaea: 15 minute
Kahului to Hana: 2 hours and 40 minutes
Kahului to Kihei: 15 minutes
Kahului to Wailea: 30 minutes
Kahului to Makena: 35 minutes
Kahului to the summit of Mount Haleakala:
    1 hour and 30 minutes
Kahului to Iao Valley State Park: 15 minutes

Kaanapali to Lahaina: 5 minutes
Kaanapali to Kapalua: 10 minutes
Kaanapali to Maalaea: 30 minutes
Kaanapali to Hana: 3 hours and 25 minutes
Kaanapali to Kihei: 40 minutes
Kaanapali to Wailea: 55 minutes
Kaanapali to Makena: 1 hour
Kaanapali to the summit of Mount Haleakala:
    2 hours and 15 minutes
Kaanapali to Kahului: 45 minutes
Kaanapali to Iao Valley State Park: 45 minutes

Kihei to Lahaina: 35 minutes
Kihei to Kaanapali: 40 minutes
Kihei to Maalaea: 10 minutes
Kihei to Hana: 2 hours and 55 minutes
Kihei to Wailea: 15 minutes
Kihei to Makena: 20 minutes
Kihei to the summit of Mount Haleakala: 1 hour
    and 45 minutes
Kihei to Kahului: 15 minutes
Kihei to Iao Valley State Park: 25 minutes

Kapalua to Lahaina: 15 minutes
Kapalua to Kaanapali: 10 minutes
Kapalua to Maalaea: 40 minutes
Kapalua to Hana: 3 hours and 35 minutes
Kapalua to Kihei: 50 minutes
Kapalua to Wailea: 1 hour and 5 minutes
Kapalua to Makena: 1 hour and 10 minutes
Kapalua to the Mount Haleakala summit:
    2 hours and 25 minutes
Kapalua to Kahului: 55 minutes
Kapalua to Iao Valley State Park: 55 minutes

Lahaina to Kaanapali: 5 minutes
Lahaina to Kapalua: 15 minutes
Lahaina to Maalaea: 25 minutes
Lahaina to Hana: 3 hours and 20 minutes
Lahaina to Kihei: 35 minutes
Lahaina to Wailea: 50 minutes
Lahaina to Makena: 55 minutes
Lahaina to the Mount Haleakala summit:
    2 hours and 10 minutes
Lahaina to Kahului: 40 minutes
Lahaina to Iao Valley State Park: 40 minutes

Wailea to Lahaina: 50 minutes
Wailea to Kaanapali: 55 minutes
Wailea to Kapalua: 1 hour and 5 minutes
Wailea to Maalaea: 25 minutes
Wailea to the Hana Lava Tube (a.k.a. Kaeleku Caverns): 3 hours
Wailea to Hana: 3 hours and 10 minutes
Wailea to Kihei: 15 minutes
Wailea to Makena: 5 minutes
Wailea to the summit of Haleakala: 2 hours
Wailea to Kahului: 30 minutes
Wailea to Iao Valley State Park: 40 minutes

IMPORTANT TRAVEL INFORMATION

- Before driving to any of the sights described in this book, be sure to *start with a full tank of gas!*
- Also, even if you have a GPS unit, be sure to also have a map. GPS units are *not* infallible!
- In the event of an emergency, be sure your cell phone is charged.
- However, in some rural areas you may not have cell phone service.
- Loss of cell phone service could be especially true on the south side of Haleakala.
- Also, have your roadside assistance phone numbers available should the need arise.
- These emergency phone numbers are generally on your rental car's keyring.
- Above all, watch for careless drivers and be sure that you are not one yourself.
- Again, users of this book assume full responsibility for their actions while using this book. Thus, they will hold the author harmless for any accidents, injuries, lawsuits, or fatalities.
- The author of this book assumes no responsibility for any of your actions.

HAWAIIAN ISLAND DIRECTIONS

- Note: you will hear local islanders using *mauka* and *makai* when giving directions.
- Islanders do not use the terms inland and seaward.
- *Mauka* is used for *inland*.
- *Makai* for *seaward*.

THE FOLLOWING WEBSITES WILL ALLOW YOU TO SPEAK LIKE AN ISLANDER

To hear *Maui place* names pronounced, go to:

http://hawaiian-word.com/hawaii-place-names

To hear *Hawaiian words* pronounced, go to

http://hawaiian-words.com/common

APPROXIMATE TRAVEL TIMES

MILEAGE CHART

Mileage from Kahului Airport to:

Destination	Mileage	Time
Haleakala	37mi	45min
Hana	52mi	2hrs
Ka'anapali	30mi.	50min
Kapalua	36mi.	60min
Kihei	9mi.	20min
Kula	18mi.	35min
Lahaina	27mi.	35min
Makena	19mi.	35min
Wailea	18mi.	23min
Wailuku	6mi.	10min

- NOTE: This table of travel times does not include any delays due to traffic.
- Give yourself more time than is indicated in the above table.

Maui and Haleakala Timeline

2 million years ago: The West Maui volcano emerged from the sea to from a single island.
900,000 years ago: Haleakala rose above the ocean. Later, lava flowing from Haleakala merged with the West Maui Volcano to form one island.
400-500 A.D.: The first Polynesians arrived, probably from the Marquesan Islands.
1100 or 1200: A second wave of settlers arrived from Tahiti.
1500: Chief Piilani came to power and started construction of the King's Highway.
1768 or 1773: Queen Ka'ahu-manu was born in a cave on Hana. She is noted for her strong will and independence. She was also the favorite wife of King Kamehameha I.
1778: Captain James Cook *saw* Maui on his return voyage to the Hawaiian Islands.
1786: The first *non-Hawaiian* landed on Maui when French explorer Admiral La Perouse landed at the bay later named for him.
1790: American Captain Simon Metcalfe committed the Olowalu Massacre, killing 100 Hawaiians. In this same year, Kamehameha I from the Big Island conquered Maui when he defeated the army of Maui's King Kahekili in a bloody battle in Iao Valley.
1795: Using Western arms, Kamehameha I wins a decisive confrontation on Oahu. Except for Kauai (which he tries to invade in 1796 and 1804), this completed his military conquest of the Islands.
1803: Kamehameha I moved his royal court to Lahaina and built the Brick Palace.
1810: The chief of Kauai acknowledges Kamehameha's rule, giving him control over Kauai and Niihau. Kamehameha becomes known as King Kamehameha I. He rules the unified Kingdom of Hawaii with an iron hand.
1819: King Kamehameha I died. The first whaling ship docks at Lahaina.
1820: The first missionaries arrived in the Hawaiian islands.
1823: Christian missionaries build the Congregational Church at Lahaina.
1824: King Kamehameha II and his wife *die of measles* in England.
1828: First recorded *non-Hawaiians* ascent of Haleakala.
1832: The Lahainaluna Seminary, the oldest school west of the Rockies, was established.
1834: The Baldwin House, Lahaina's oldest standing building was built.
1840: Kamehameha III officially decreeds Hawaii's first constitution from Lahaina. The United States identifies Pearl Harbor as a potential Naval Base.
1846: Maui's first Catholic mission was established.
1849: Kamehameha III turns Hawaii into a constitutional monarchy, and the United States, France, and Great Britain recognize Hawaii as an independent country.
1849-51: In Hana, George Wilfong started Maui's first sugarcane plantation.
1874: David Kalakaua begins his reign as the "Merrie Monarch."
1875: United States and Hawaii sign a treaty of reciprocity, assuring Hawaii a duty-free market for sugar in the United States.
1852: Western diseases depletes local labor and mass worker immigration begins.
1862: The Pioneer sugar mill was built at Lahaina.
1873: Lahaina's famous banyan tree was planted.
1876: Samuel Alexander and Henry Baldwin began construction on the Hamakua Ditch to bring irrigation water to their sugarcane fields in central Maui.
1883: Six sugar plantations were in operation in Hana: The Kaeleku Sugar Co., Hamoa Agri. Co., Kawaipapa Agri. Co., Hana Sugar Co., Reciprocity Sugar Co., and Haneoo Agri. Co.
1891: King Kalakaua dies, he is succeeded by his sister, Queen Liliuokalani, the last Hawaiian monarch.
1893: After a brief two-year reign, Liliuokalani is removed from the throne by American business interests. Liliuokalani is imprisoned in Iolani Palace for nearly eight months.
1901: The Pioneer Inn opened in Lahaina. It remained West Maui's major hotel for more than 50 years.
1916: Haleakala was included as a part of Hawaii National Park
- *Continued on the next page.*

1926-27: The road to Hana was built, making it possible for trucks and cars to travel from Hana to Kahului
1929: Inter-Island Airways, later to become Hawaiian Airlines, landed its first plane on Maui.
1935: First permanent ranger stationed at Haleakala.
1936: Haleakala Visitors Center was built, and the road to the summit was constructed by the Civilian Conservation Core (CCC). The Work Project Administration (WPA) is dedicated.
1937: CCC completed construction of the Kapalaoa, Holua, and Paliko cabins on Haleakala.
1900-1940: Hana is a bustling town with a population of 3,500, having two movie theaters, 15 different stores, three barber shops, a pool hall, a jail, and several restaurants.
1944: Paul Fagan started the Hana Ranch on 14,000 acres of land and a herd of Herford cattle he brought from Molokai.
1946: April 1, a tsunami struck the Hana coast killing 12 people. Devastation was far worse at Hilo on the Big Island.
1946: The last Hana sugar plantation closed, ending the sugar industry in Hana. Paul Fagan, a retired entrepreneur from San Francisco, started the Ka-'uiki Inn, now known as the Hotel Hana.
1959: Hawaii became the 50th state.
1959: Kaanapali was developed as Hawaii's first planned resort.
1961: Haleakala, once combined with Volcanoes National Park on the Island of Hawaii, became a *separate* national park. The Haleakala National Park headquarters building is completed.
1962: The Haleakala summit building was completed.
1962: Lahaina was designated a National Historic Landmark.
1969: Kipahulu Valley was added to Haleakala National Park.
1970: Paul Fagan died. In his memory, a cross made of basalt was erected on a hill overlooking Hana Town
1974: Charles Lindbergh, the famous aviator, spent his last days in his home in Hana before dying. He was buried in a quiet Maui cemetery.
1976: Haleakala National Park obtained a wilderness designation for 19,270 acres and constructed a boundary fence to control feral goats.
1984: The Hotel Hana-Maui and Hana Ranch were bought by the Rosewood Corporation of Dallas, Texas.
1988: The Crater District boundary fence was completed
1989: The Hotel Hana-Maui and Hana Ranch were bought by the Keola Hana-Maui, an international investment group made up of Japanese, British and local investors.
1993: About one million visitors go to Haleakala's summit, and about 500,000 visitors go to the Kipahulu portion of Haleakala.
2000: Meridian Financial Resources, an organization of Chicago-area investors, takes control of the Hotel Hana-Maui and Hana Ranch from Keola Hana Maui.
2001: Meridian Financial Resources sold the 4,500 acre Hana Ranch to Hana Acquisition Partners, which was largely financed by the Ronald Getty Trust and Susan and Roy O'Connor.
2001: Passport Resorts of San Francisco bought the Hotel Hana Maui and associated Hana Town Center businesses. Passport Resorts also owns the Post Ranch Inn in Big Sur and the Jean-Michael Cousteau Resort in Fiji.
2016: By December 2016, after sustaining a $30 million loss in 2015, and with the future not looking any better, the Hawaiian Commercial and Sugar Co. closed their mill on Maui, which was the last sugar mill in Hawaii.
2017: In February 2017, Pacific Biodiesel Technologies began its farming demonstration to grow biofuel crops, including sunflowers, in Maui's central valley.
This project is the largest biofuel crop project in the state of Hawaii.

WEBSITES FOR NEWS AND INFORMATION

The Maui News: Maui's daily newspaper: **mauinews.com**

The Lahaina News: Weekly newspaper with news for West Maui: **lahainanews.com**

Maui Weather Today: Three times a week, a Maui weatherman gives weather forecasts and comments on life on Maui to the Internet. This site has several links to satellite shots of the islands: **hawaiiweathertoday.com**

Maui Now: Up-to-the minute coverage of what's happening on Maui. Plus: events, reviews, weather, sports, business, and more: **mauinow.com**

Maui Time: Maui News covering the best of Maui, things to do in Maui, restaurant reviews, places to eat, entertainment, politics and more: **mauitime.com**

News History of Hawaii: Summarizes the history of the island of Maui:
gohawaii.com/maui/about/history

Maui Museum: mauimuseum.org

Best of Maui Guide: Lahaina Through the Ages: **bestofmauiguide.com/lahainahistory**

Live Cameras

Maui Webcams: Webcam shows scenic viewpoints of Maui: **mauihawaii.org/webcams.htm**

Maui Live Surf Cam - Kaanapali - Lahaina - Wailea - Kapalua:

livesurfcamhawaii.com/maui/maui.htm

MAP OF THE MAUI'S VOLCANOES

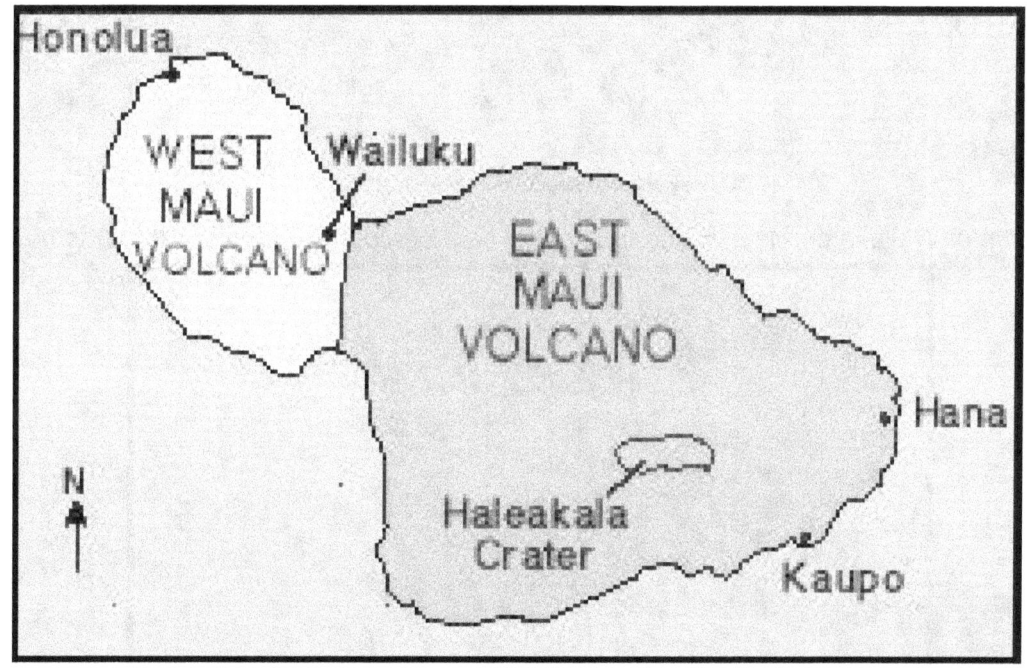

Courtesy of the USGS

BRIEF MAUI OVERVIEW

- Maui is composed of two sections: The West Maui Volcano on the northwest end of the island and Haleakala toward the southeastern portion of the island. See the map below.

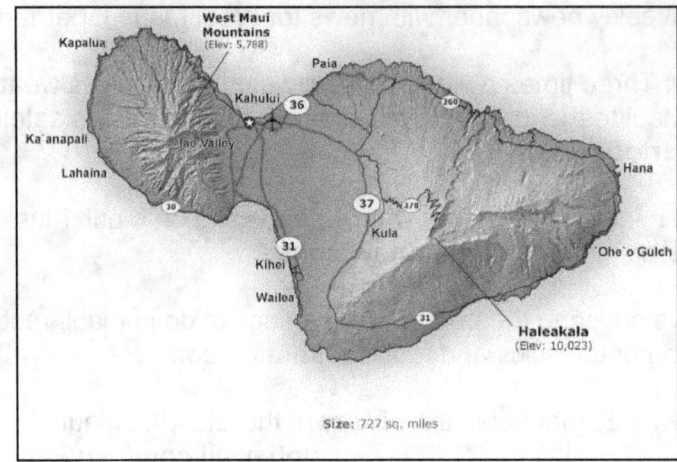

Modified from Hawaii-guide.com

- These two areas are separated by the narrow Maui Isthmus.
- Maui is 48 miles long and 26 miles wide, or six miles wide at its isthmus.
- In terms of area, Maui has an area of 728 square miles.
- Flying into Maui, you will land on the east side of Maui at the Kahului Airport in Kahului
- The airport code for this airport is OGG, named for Captain Jimmy Hogg, a Hawaiian Airlines pilot and aviation pioneer. *For more details Google Captain Jimmy Hoog.*
- Maui's largest city is Kahului (population 20,000).
- Maui is also the second-largest Hawaiian island. *See the map below.*

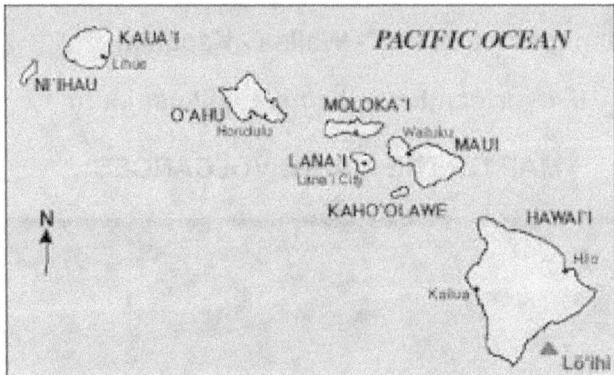

Courtesy of the United States Geological Survey

- The town of Wailuku on Maui is also the county seat of Maui *County. See the map below.*

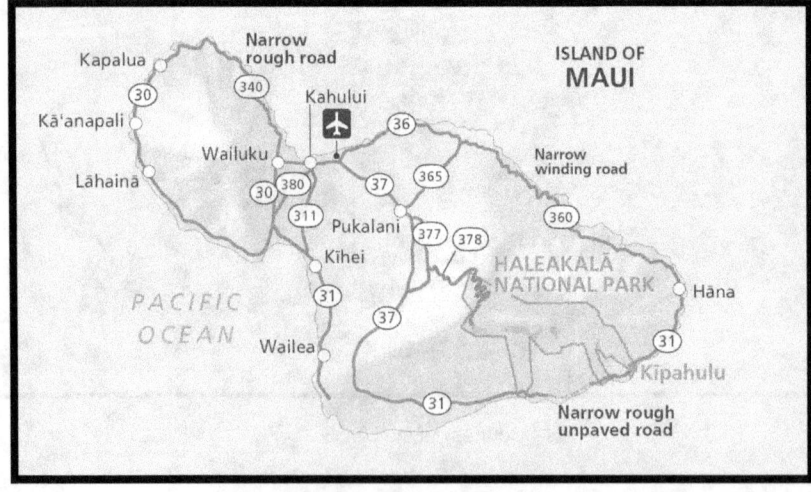

GEOGRAPHY OF THE HAWAIIAN ISLANDS AND MAUI

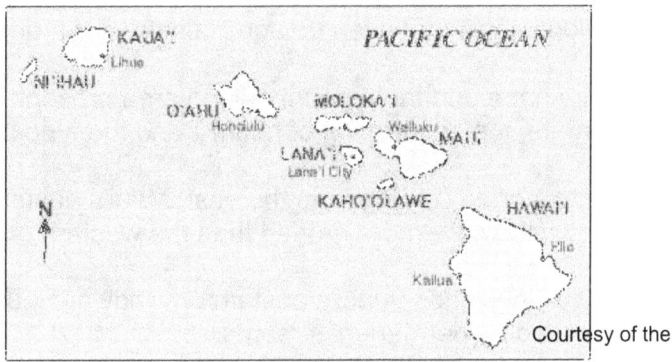

Courtesy of the USGS

- The Hawaiian islands, shown in the above map, are one of the most isolated islands in the world.
- These islands are approximately 2400 miles southwest of California, 4,100 miles from Japan, and 5280 miles from the Philippines.
- Thus, the Hawaiian Islands lie between latitudes 19 degrees and 29 degrees North and longitudes 154 and 179 degrees West.
- The Hawaiian archipelago consists of 132 islands, islets, sand cays, and reefs.
- However, most of the total land area consists of five major islands: Hawaii, Maui, Oahu, Kauai, and Molokai.
- In addition to these five islands, three smaller islands: Lanai, Niihau, and Kahoolawe complete the main islands of this island chain.
- Maui, encompassing 728 square miles, is the second largest of the main Hawaiian Islands,.
- But, the island of Hawai'i, called the "Big Island," contains more than twice as much land as the other seven islands combined.
- Due to their isolated location, these islands were among the last areas in the world to be discovered and populated.
- The greatest distance between any two of the larger Hawaiian Islands is the eighty miles from Kauai to Oahu. Distances between adjacent islands averages twenty-five miles or less.
- Thus, except for certain wide channels, the earliest Hawaiians were able to easily sail between these islands.
- The Hawaiian Islands are volcanic in origin, having formed over the Hawaiian Island hotspot, which will be discussed on page 18.
- Kauai is the oldest island, and the Island of Hawaii, at the southeast end of the island chain, is the youngest. *See the above map.*
- Maui formed from the joining of two shield volcanoes, the West Maui Mountain and Haleakala. *See the map below.*

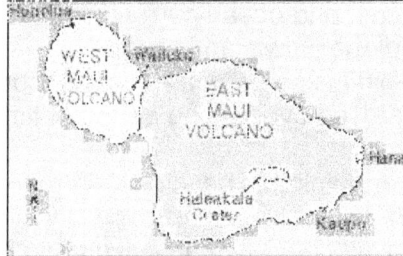

Courtesy of USGS

- The low-lying Maui Isthmus lies between these two volcanoes.
- The West Maui Volcano, the older of the two volcanoes, has eroded such that, today, its highest peak, *Puu Kukui*, is only 5,788 feet tall.
- Haleakala, Maui's younger volcano, has an elevation of 10,023 feet.
- *But,* when measured from the seafloor, it has a height of 29,704 feet, which is higher than Mount Everest's 29,029 ft.
- Thus, Haleakala is the largest dormant volcano in the world. *Continued on the next page.*

- The last volcanic eruption on Maui was between 1480 and 1600 when lava was extruded from Haleakala's Southwest Rift Zone. More details will be given on page 25 and 29.
- Thus, this historic eruption shows that even though Haleakala is a dormant volcano, it is not extinct.
- Due to variations in their shape and the amount of land area present, each of the Hawaiian Islands differ in their climate, the occurrence of rainfall, and the type and distribution of vegetation.
- In regards to Maui, due to storms coming from the east, Maui's eastern slopes receive more rainfall, and, consequently, they are more eroded than the western (windward) side of the island.
- Thus, like the other Hawaiian Islands, Maui's eastern or windward side is cut by steep valleys and cliffs, often being eroded by permanent streams.
- Because of this difference in rainfall, the *warmer and drier leeward sides* of the islands are: less eroded, are characterized by grassland and brush, have shallower trough-like valleys, have coastal plains with flat sand or cobble beaches, and occasional coral reefs.
- Also, because the western side of Maui is warmer and has a drier climate, this region, like on Hawaii and Oahu, was developed as a resort area.
- In terms of vegetation, the predominantly wet and cool eastern portion of Maui is covered with trees, abundant grass, and other types of plants.
- Maui's "green" eastern side can be seen in the satellite photo below.

Courtesy of NASA

- However, prior to the arrival of Europeans, Maui was more heavily forested.
- One cause of this deforestation was the European demand for fragrant sandalwood.
- In terms of vegetation: wind, birds, and ocean currents introduced a variety of plants to the mild, subtropical western side of the Hawaiian Islands.
- Thus, the "native" plants and animals seen on the islands today are: 1) the descendants of those plants and animals that arrived over a long period of time, **or** 2) were introduced by the early Hawaiians.
- Since then, many other plants and animals have been introduced by the early Europeans, **or** more recently by other humans.
- Unfortunately, not all of these plants and animals have been beneficial. For example, the tree frogs on the Big Island are a serious pest.
- Due to the varied geography of the Hawaiian Islands, a variety of environments exist.
- Consequently, some of Hawaii's flora and fauna are only found in specific environments.
- Two examples of specific environments would be the flora at the top of Mount Kaala on Oahu, or the flora in Maui's Kipahulu Valley.
- *Continued on the next page.*

COMMUNITY DISTRICTS OF MAUI

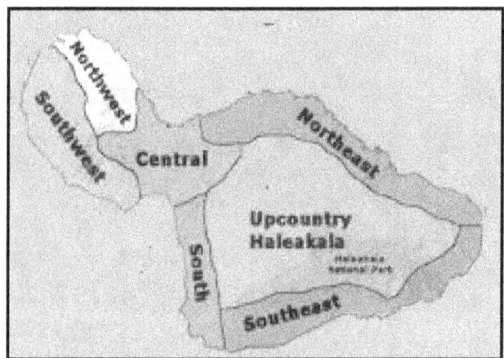

Courtesy of Maui Community Plan

Refer to the above map for the location of Maui's Community Districts described below.

Southwest Maui:
- Southwest Maui is where ancient Hawaiians of royalty would gather.
- Today, this region has a beautiful coastline with luxury resorts and shopping venues.
- Southwest Maui is mostly sunny. It is also the site of the famous whaling town of Lahaina.
- Kaanapali and Kapalua, with light colored sandy beaches, are the location of the main resorts in this area.

South Maui
- The South Maui region has miles of light colored sandy beaches, but swimming can be dangerous.
- This area is Maui's sunniest, driest, and most popular area, with a number of high-end resorts.
- Wailea, located in south Maui, is a planned resort area that was once dry and barren, but due to an irrigation project, it is now the site of some of Maui's most exclusive resorts.
- Wailea has nearly 2 miles of five crescent beaches connected with a one and a half mile botanical walking path.
- Swimming and snorkeling is good here, but there are hazards (Clark, 1989).

Central Maui
- Has many historical Native Hawaiian sites and small towns. It is also the location of Maui's airport at Kahului.
- To avoid traffic and the cost of the large resorts, many local people live in this region.
- This region has shopping malls and a number of big box stores such as Costco and Wal-Mart.
- Wailuku town, the county seat of Maui County, is also located here.
- Central Maui's isthmus was once the location of over 36,000 acres of sugar cane.

Upcountry
- Upcountry Maui is located on the green slopes of Haleakala.
- Due to its higher elevation, this area is often much cooler than the lower regions of Central Maui.
- Upcountry is where: ranchers raise cattle, flowers are grown, and where a variety of vegetables are grown (such as the popular Maui onion).

Northeast Maui
- Northeast Maui begins at the old sugar plantation town of Paia.
- Many small stores and cafes are found there.
- The road from Paia leads to Hookipa, one of the world's best wind surfing beaches.
- Further southward is the famous "Road to Hana"; with its waterfalls, tropical native Hawaiian rainforest vegetation, 56 one lane bridges, and countless curves.
- This rewarding drive will take you to: Hana, the interesting Hasegawa General Store, and then to the Seven Sacred Pools.

ELEVATION MAP OF MAUI

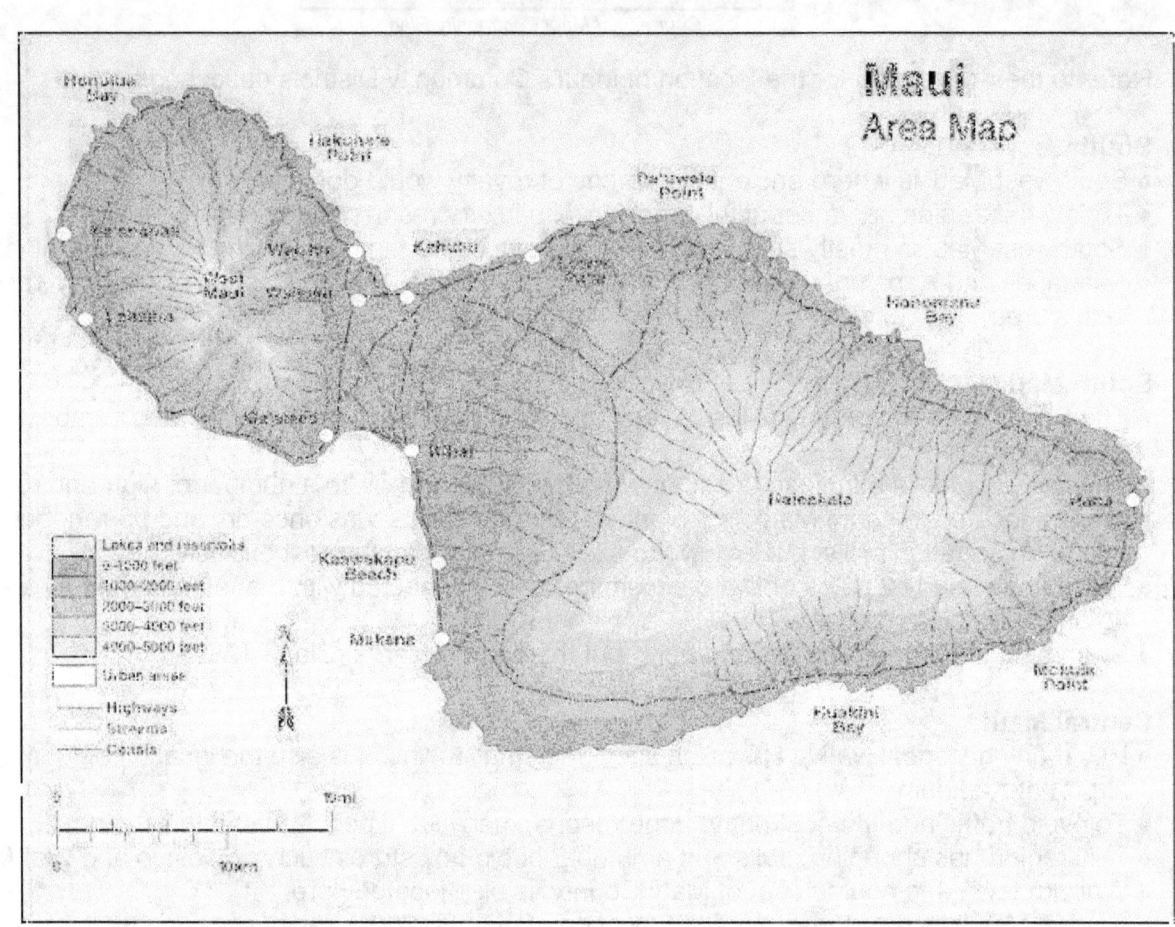

CLIMATE OF MAUI

- Maui has a moderate, tropical climate with mild winters and warm, dry summers..
- **Summer months**: May through September, very warm, with sunny days.
- Hottest months: August and September; temperatures rise to 89°F.; averaging 79° F.
- **Winter months:** mid-October through to April.
- Coldest months: January and February; temperatures from 80°F to 64°F; averaging 71°F.
- The wettest periods are from mid-November through late March.
- The distribution of Maui's rainfall can be seen in the rainfall map below.

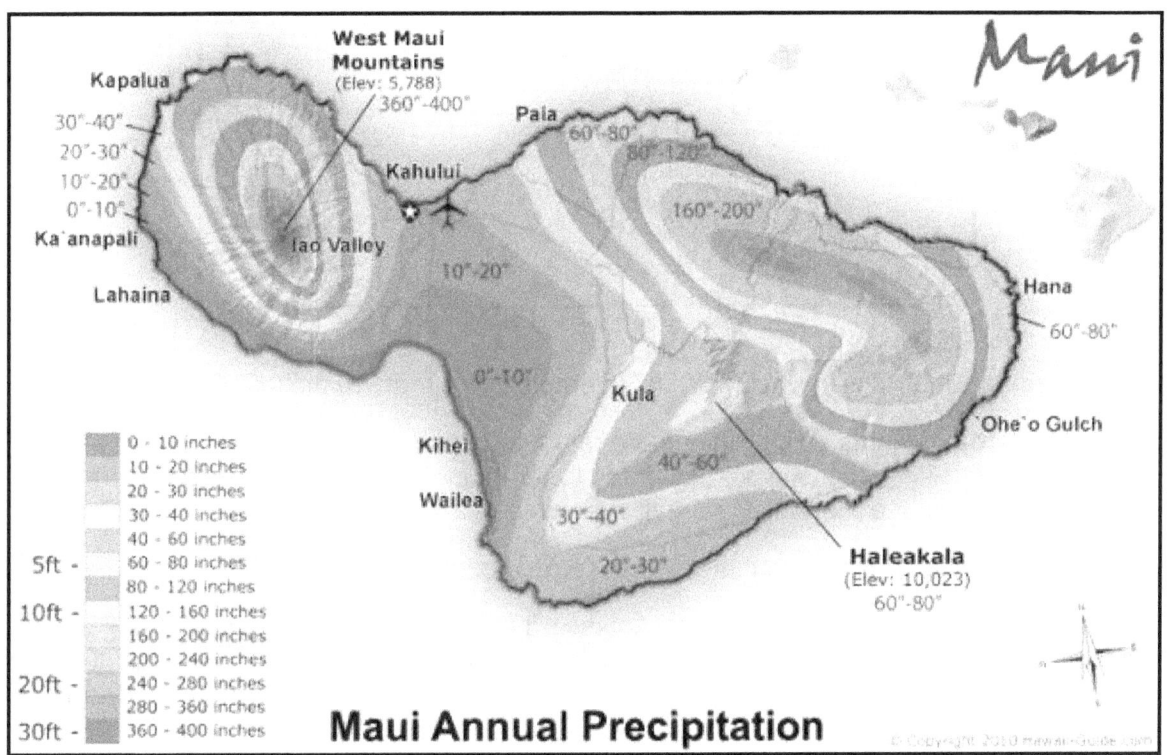

The above map used with permission from Hawaii Guide.com

DISTRIBUTION OF MAUI VEGETATION VERSUS RAINFALL

- As the image below shows, the distribution in Maui's vegetation changes dramatically as the amount of rainfall changes.

Image modified from National Park 3-d image.

- *Continued on the next page.*

GEOGRAPHIC DISTIBUTION OF MAUI'S RAIN FALL

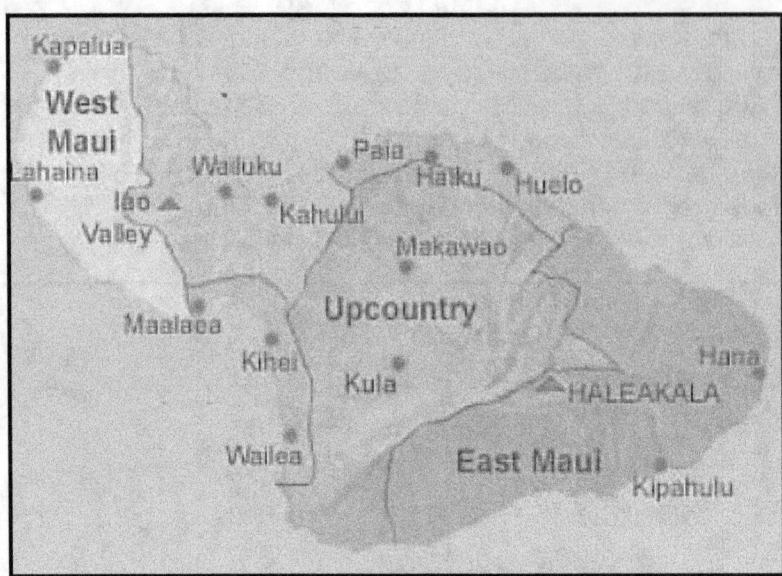

Modified from unidentified source

1.) **Driest areas:** Kihei, Wailea and Makena, averaging less than 10 inches of rain each year.

2.) **Northwest coast:** Kaanapali, slightly more rain than southern areas. From Kapalua northward rainfall increases in coastal areas to 30-40 inches per year. Towards the West Maui Volcano and the Iao Valley, rainfall increases to 360 to 400 inches/year (400"=33ft).

3.) **Inland in the Upcountry:** around Kula this region receives 30-40" a year.

4.) **Northeast coastline:** rainfall increases eastward and inland. Paia receives 40-60" a year. Areas east of Paia may receive up to 160 to 200 " per year. Hana receives 80" annually, while at Oheo Gulch, the rainfall increases to 80-120 inches (120" = 10 feet).

TRADE WINDS: are usually at their lowest frequency in September and October.
- account for 70% of all wind in Hawaii; they are the most common winds over Hawaii.
- blow from the NE to ENE at about 5-15 miles per hour; they moderate Maui's climate.
- in the summer, they prevail more than 90% of the time, sometimes lasting an entire month.
- in the winter, January through March, trade winds may occur only 40% to 60% of the time.

KONA WINDS: are stormy, rain-bearing winds that blow over the islands from the SW or SSW, that is, they blow in the opposite direction of trade winds.
- usually don't last for more than a day or so.,
- bring the gases emitted from Kilauea's eruptions (the vog) up to other Hawaiian Islands.

SUMMARY OF HALEAKALA'S WEATHER

Annual temperatures:
- When driving to Haleakala's summit, the temperature decreases about 3° F. per 1000 feet of elevation. Thus, you often need to bring a jacket or sweatshirt.
- Average *annual* temperature at the park headquarters (elevation 7000 feet) is 53° F.
- Average *annual* temperature at the 10,023 foot summit is even colder.
- At the park headquarters, highest recorded temperature was 80° F. (October 1973).
- Lowest recorded temperature was 30° F.: April 1969, January 1982, January 1990.
- Wind? December 1990; a wind indicator near the summit broke at a wind speed of 128 mph.

Average annual rainfall:
- At the park headquarters (7000 feet) the average annual rainfall is 53 inches.
- At the 10,023 foot summit, average annual rainfall decreases to 40 inches.
- Maximum amount of rainfall in a 24 hour period; 18.5 inches occurred in January 1980.

DROWNINGS ON MAUI

- According to the state Department of Health, ocean drownings are the second leading cause of fatal injuries in Maui County.
- Maui County reported visitors accounted for 101 out of the 138 ocean drownings between 2006 and 2015.

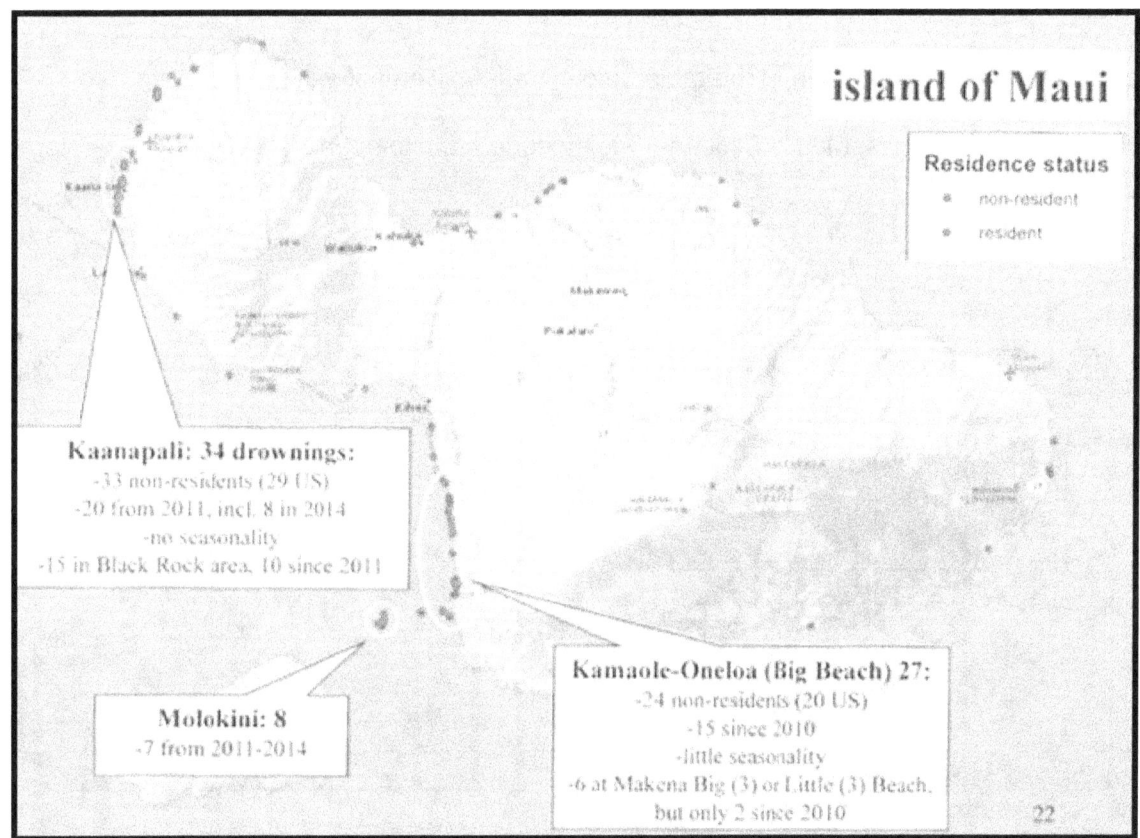

Courtesy of the Hawaii Department of Health

BEACH SAFETY WEBSITES

http://hawaiibeachsafety.com/maui

http://mauiguidebook.com/things-to-do/the-basics/ocean-conditions/

WEBSITE OF SHARK ATTACKS ON MAUI

http://dlnr.hawaii.gov/sharks/shark-incidents/incidents-list

SIMPLIFIED GEOLOGIC TIME SCALE

The lava flows on Maui's are Holocene, Pleistocene and Pliocene in age. Thus, the time scale below has been enlarged to show the subdivisions (epochs) of the Tertiary and Quaternary Periods.

The ages of these units shown in the table below are in millions of years.

As this time scale shows, Maui's Pliocene, Pleistocene and Holocene lava flows are geologically very young.

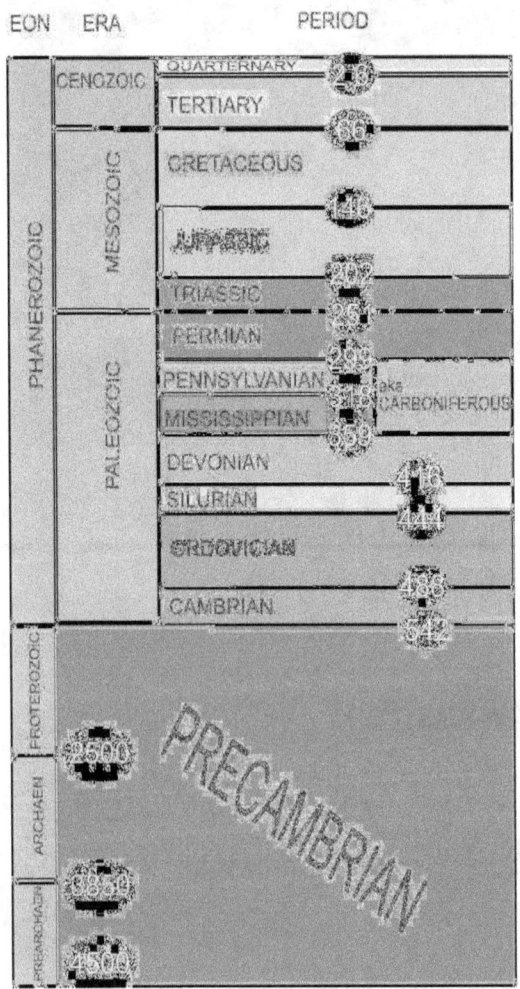
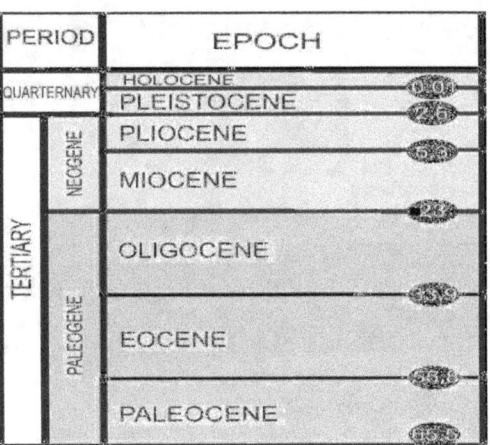

THE EARTH IS ABOUT 4.6 BILLION YEARS OLD

THE ORIGIN OF THE HAWAIIAN ISLANDS

- The Hawaiian Islands are composed of a linear chain of islands approximately 125 volcanic islands and submarine volcanoes extending northwestward from the Island of Hawaii to the Aleutian Trench. *See the map to the lower left.*

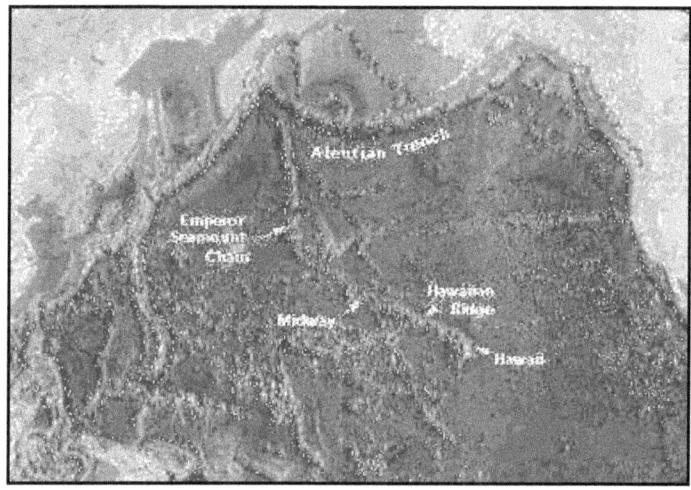

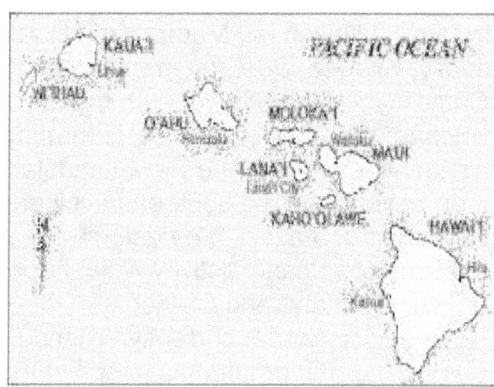

Courtesy of the United States Geological Survey

- The Hawaiian Island portion of this chain consists of 8 main islands. *See the map to the upper right.*
- The Hawaiian Islands formed as a series of volcanoes rose from the sea floor. *See the illustration to the right.*
- These islands become progressively older going from the Island of Hawaii (the Big Island) towards the Aleutians Trench. *Again, see the map to the right.*
- Thus, Kauai is the oldest island in the chain, having formed 5.6 million years ago, and the Island of Hawaii is the youngest. *Again, refer the map to the right.*
- The *entire* Pacific island chain formed as the North Pacific Plate move over the Hawaiian Island Hotspot. *See the illustration below.*
- This hotspot formed as a plume of hot magma (molten rock) rose from deep within the Earth to form volcanoes on the overlying northwest moving Pacific Plate. *See the diagram below.*

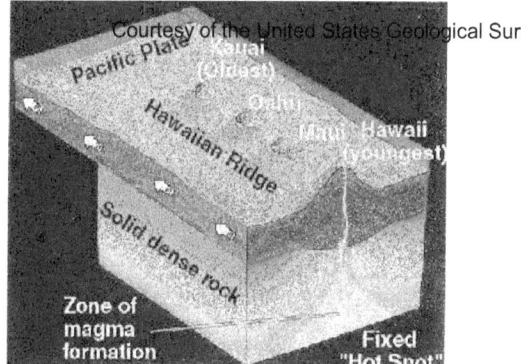

Courtesy of the U.S. Geological Survey

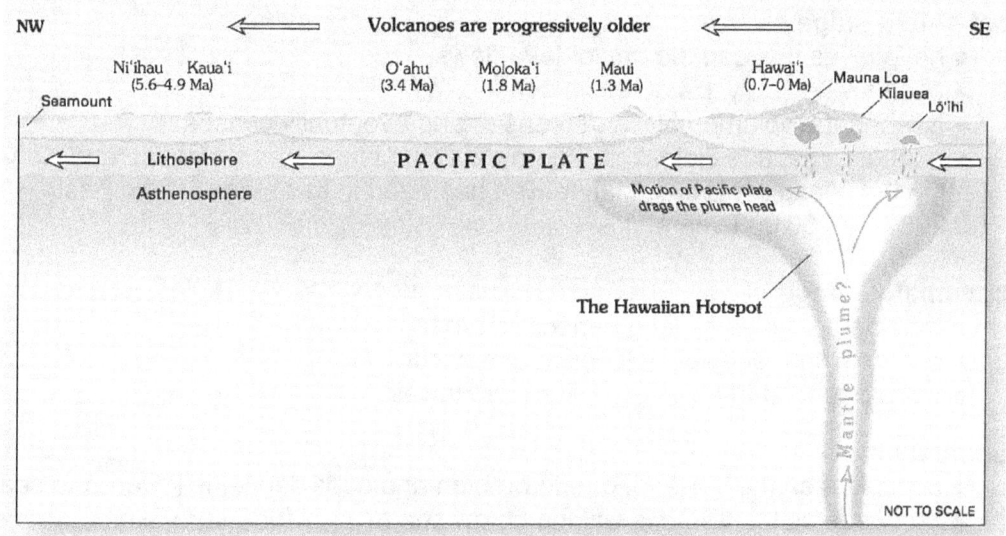

Courtesy of Joel E. Robinson, USGS

CONTINUED ON THE NEXT PAGE

- Thus, as the Pacific Plate moved to the northwest, each volcano was carried away from the Hawaiian Island Hot Spot.
- And, one after another, as they were cut off from their source of lava, these volcanoes became dormant and then extinct. For example, on the Island of Hawaii: Kohala was the first volcano to form, followed by, Mauna Kea, Hualalai, Mauna Loa, and finally, Kilauea. *See, the map to the right.*
- Currently, on the sea floor southwest of the Big Island, a *new* volcano, the Loihi seamount, is forming,
- As the volcanoes in the Hawaiian Island chain passed over, and then beyond the hot spot, *each* volcano progressed through a sequence of stages. *See below.*
- In terms of volcanic activity, volcanoes can be described as: Active, Dormant, and Extinct.
    Active: currently erupting.
    Dormant: has past historic eruptions and expected to erupt again.
    Extinct: No historic eruptions and not expected to erupt again
- Note: Haleakala on Maui had its most recent eruption between 1480 and the 1600's.
- Thus, Haleakala is not an extinct volcano. Rather, it is in a dormant stage.

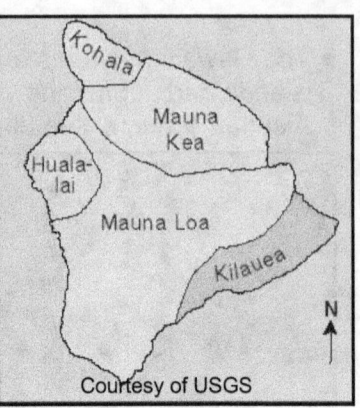
Courtesy of USGS

GENERALIZED STAGES IN THE DEVELOPMENT OF A HAWAIIAN ISLAND VOLCANO

1. Pre-shield stage:
- when the volcano is slowly building up from the sea floor.
- lasts about 100,000 years. Example: Loihi

2. Shield stage:
- when the classic shield shaped volcano forms.
- when large amounts of basaltic lava passively extruded.
- forms a volcano with broad sloping sides.
- when 95% of the volcano's volume forms.
- when a large summit calderas forms.
- when volcanoes have flank eruptions of fluid basaltic lava.
- lasts for about 500,000 years.
- examples: Mauna Loa, Kilauea and Haleakala.

3. Post-shield stage:
- have lavas that cap the earlier lava flows.
- occurs when only 1% of the volcano forms.
- is when the volume of lava decreases and eventually ceases.
- examples include Haleakala (Sinton 2003), Mauna Kea, Hualalai, and Kohala. On the Island of Hawaii, Mauna Kea has been in this stage for the past 200,000-250,000 years.

4. Rejuvenated stage:
- may be preceded by long periods of erosion.
- occurs when renewed volcanism can occur.
- not present on the Islands of Maui or Hawaii.

5. Subsidence:
- occurs when the volcano, due to erosion and subsidence, is lowered to sea level.
- proceeds as the island is sinking below sea level to form coral atolls.
- produces flat topped mountains, called guyots, when the volcano is fully submerged.

GEOLOGY OF MAUI

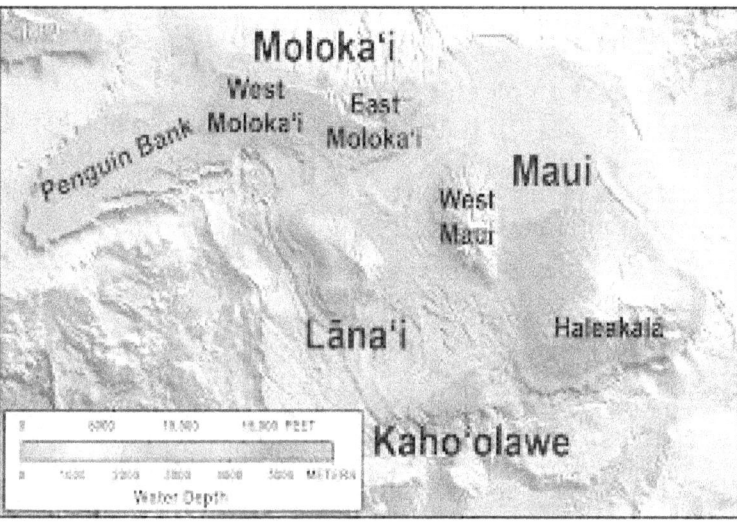

Courtesy of USGS

- 1.2 million years ago, the islands of Maui, Molokai, Lanai, and Kahoolawe were, connected as a single landmass known as *Maui Nui,* literally *"Big Maui."* See the illustration to the right.
- Maui Nui encompassed about 5,640 square miles, or, was 50% larger than today's Big Island.
- However, the earth's crust could not support the weight of these huge island volcanoes, and the crust subsided at rate of 3 mm per year (0.1 inch per year).
- Consequently, within the past 1.2 million years, *Maui Nui* began to breakup as the **Penguin Bank** and the **West Molokai volcano** broke apart. *See the above map.*
- The Penguin Bank was a broad shoal area west of Molokai. It originated as a volcano that completed its shield-building stage about 2.2 million years ago, when it was briefly connected to Oahu.
- Then, about 700,000 years ago, a permanent embayment separated **East Molokai** from **West Maui**, but these two volcanoes remained connected via Lanai.
- Then, about 600,000 years ago, the low area between **East Molokai** and **Lanai** submerged.
- The breakup between **Kahoolawe** and **Maui** probably took place between 200,000 and 150,000years ago.
- However, as recently as about 18,000 years ago, Maui, Lanai, and Molokai were intermittently connected.
- Then 12,000 to 10,000 years ago, when the Pleistocene glaciers finally melted, sea level rose.
- As a result, the low lying areas between these four islands was submerged, *isolating these different volcanoes as the islands of Maui, Lanai, Molokai, and Kahoolawe.* See the illustration to the right.

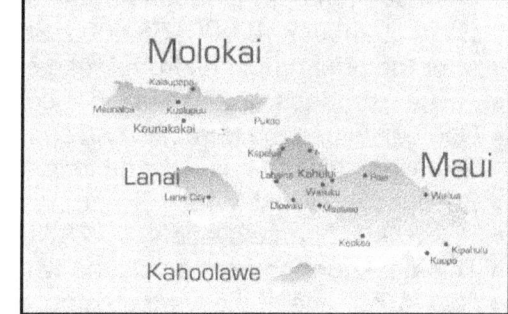

Courtesy of USGS

VOLCANOES ON THE ISLAND OF MAUI

- The Island of Maui is the second largest of the Hawaiian Islands, and the second youngest of the main Hawaiian Islands.
- Maui, is composed of two volcanoes: the West Maui Volcano and the East Maui Volcano (*aka, Haleakala*). See the map below.

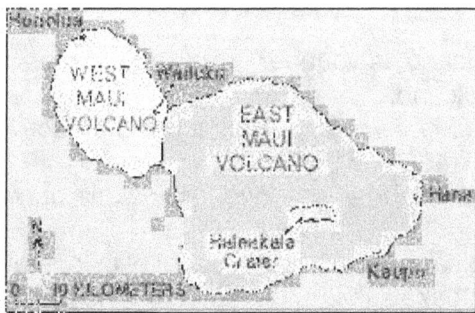

Courtesy of USGS

- *Continued on the next page.*

ORIGIN OF THE WEST MAUI VOLCANO
(Source for geologic ages: Geologic Resources Inventory Report NPS)

In a *simplified* description, from the *oldest to youngest*, the West Maui Volcano, consists of:

- **The Wailuku Volcanic Series:**
  - extruded 2.0 to 1.3 million years ago during a shield building phase that formed about 97% of the volcano's volume.
  - consists of thin ropy and blocky lava flows that are mainly iron rich, olivine bearing basalts.
  - were generally extruded as quiet eruptions, with violent pyroclastic events being rare.,
  - was when a summit collapse formed a caldera almost 2 miles across.

- **Honolua Volcanic Series:**
  - extruded 1.3 to 1.1 million years ago.
  - formed an incomplete cap covering of a lighter colored, less iron rich basaltic lava.
  - had a long period of erosion following the extrusion of this lava.
  - During this time, streams eroded deep canyons in the sides of the volcano and oceanwaves eroded steep sea cliffs.

- **Lahaina Volcanic Series:**
  - extruded 0.6 to 0.3 million years ago during a period of weak volcanic activity that built four small volcanic cones.
  - near the shoreline, on its southwestern slope, this lava was slightly different in composition from the previously extruded basalt

ADDITIONAL DETAILS ON THE WEST MAUI VOLCANO (aka, Mauna Kahalawai)

- West Maui's main volcanic eruptions culminated its growth about 1.1 million years ago.
- Then, four small cinder cone eruptions occurred during the West Maui Volcano's rejuvenated stage at about 610,000, 580,000, 388,000, and 385,000 years ago.
- After the brief eruptions from these cinder cones, these cinder cones became extinct.
- Lavas extruded from these cinder cones generally only covered a few acres.
- Geologists use the phrase *"rejuvenated-stage"* for these eruptions that occur long after a Hawaiian volcano has finished its main stage of growth.
- Because the West Maui Volcano has not erupted in historic time, it is considered to be extinct.
- *Puu Kukui,* at an elevation of 5,787 feet, is the West Maui Volcano's highest peak.
- This peak formed from a volcano whose caldera eroded into what is now the *Iao Valley*.
- *Puu Kukui*, one of the wettest spots on Earth, averages 362 inches per year (or 30 feet/year).
- The West Maui Volcano has been deeply eroded by the Waihee, Waiehu, Iao, and Waikapu streams.
- Sediments deposited by these streams formed alluvial fans that helped to create the Maui Isthmus.
- In addition, *Puu Kukui* is home to many endemic plants, insects, and birds.
- However, access to this area is restricted to researchers and conservationists.

GENERAL DESCRIPTION OF THE HALEAKALA VOLCANO
(East Maui Volcano)

- Haleakala, a large shield volcano, is younger, taller, and has a smoother surface topography than the older West Maui Volcano.
- Also, unlike the extinct West Maui Volcano, Haleakala is an active volcano, with historic eruptions occurring between 1480 and 1600.
- With an elevation of 10,023 feet above sea level, Haleakala only exposes seven percent of its entire mass.
- Fifteen times more of the volcano seen above sea level is located below sea level.
- Thus, from the sea floor, Haleakala is 29,703 feet tall, which is 674 feet taller than Mount Everest.
- *Continued on the next page.*

RIFT ZONES

- The West Maui Volcano and Haleakala have three rift zones.
- **Note**: Rifts are fractures in the earth's crust through which lava and pyroclastic debris (ash, cinders, blocks and bombs) are ejected.
- Haleakala's three rift zones are: the Northwest Rift Zone, the Southwest Rift Zone, and the East Rift Zone. *See the map below.*

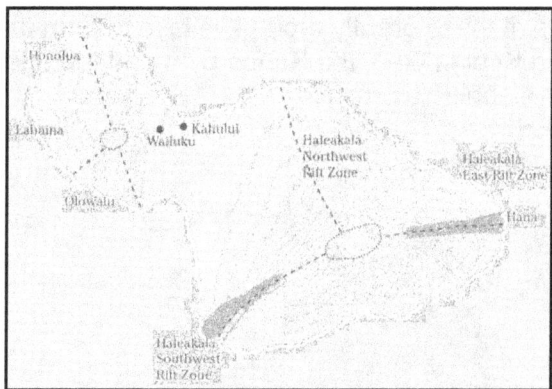

- These rift zones are quite evident on the digital elevation model of Maui below.
- Note: A digital elevation model (DEM) is a digital model or a 3D representation of a terrain's surface

Courtesy of the USGS

- At the northwest end of this image, the West Maui Volcano can also be seen.
- Because this volcano is older than Haleakala, it has been more deeply eroded.
- This erosion is easily seen as the deep canyons that radiate from its summit.

CALDERAS

- Calderas are large summit depression that are *much large* than a volcanic crater.
- Calderas form when large volumes of magma are either extruded by a volcanic eruption, or withdrawn into the subsurface, or erupted from side vents. *See diagram below.*

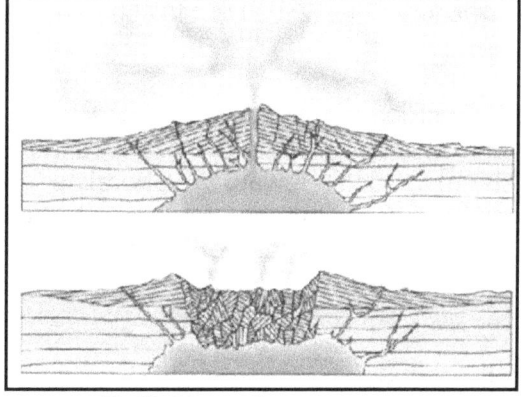

Modified from unknown source

- With the structural support for the overlying rock thus removed, the walls of the volcano slump inward to form this large summit opening, aka, the caldera. *Continued on the next page.*

- However, *Haleakala's summit basin formed differently.*
- The summit of Haleakala is an oval-shaped depression seven miles long and 2.5 miles wide that is *commonly called a crater*. However, this depression was **not** volcanically produced.
- Haleakala's summit basin formed by a long period of **stream erosion** as the headwalls of two valleys eroded headward toward their source, enlarging and deepening their respective valleys.
- Eventually, the heads of the Keanae and Kaupo Valleys almost joined when lava flooded the region between the heads of these greatly eroded valleys to form the summit basin.
- The map below shows the Koolau Gap, the Kaupo Gap and the ages of the lava flows in this summit basin, aka, mislabeled a crater.

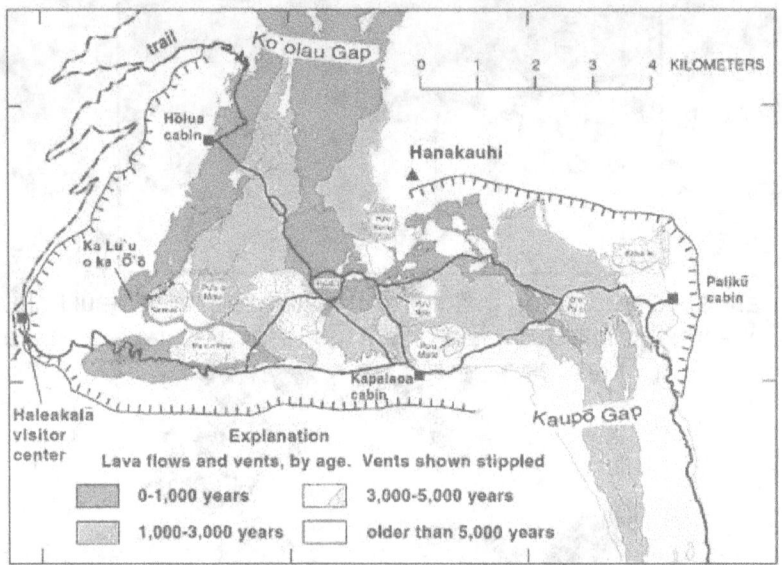

Modified from http://hvo.wr.usgs.gov/volcanoes/haleakala/cratermap_large.jpg)

VOLCANIC UNITS ON HALEAKALA
(Source for geologic ages: Geologic Resources Inventory Report NPS)

From the youngest to the oldest:

Hana Volcanics:
- extruded after a long period of erosion that eroded deep canyons later filled with these fluid Hana lava flows.
- represents renewed rift zone volcanism.
- are lavas similar in composition to the older Kula Volcanics, but were thinner and more fluid.
- included explosive eruptions that formed large cinder cones from Hana up the East Rift Zone, across the summit, and down the South Rift Zone to La Perouse Bay.
- were extruded from 0-1500 years ago to: 50,000 to 140,000 years ago

Kula Volcanics:
- extruded 930,000 to 150,000 million years ago.
- mantled the older Honomanu shield volcanics, are exposed in the walls of the "crater".
- flows are thickest at the summit and thin toward the shoreline.
- composed of hawaiite basaltic lavas, with a greater amount of ash and cinders from explosive eruptions that formed many large cinder cones along the northwest, southwest, and southeast rift zones.

Honomanu Basalt:
- extruded 1.1 to 0.97 million years ago.
- have few exposures, found mainly in deep canyons.
- composed of alternating blocky and ropy lavas *extruded during the shield building phase*.
- flowed down Haleakala's northwest slopes impounding against the West Maui Volcano to form the Maui Isthmus. **These volcanic units are shown on the maps on the next page.**

GENERALIZED GEOLOGIC MAPS OF MAUI

Modified from Sherrod et al 2003

Modified generalized geologic map of Maui compiled by various authors

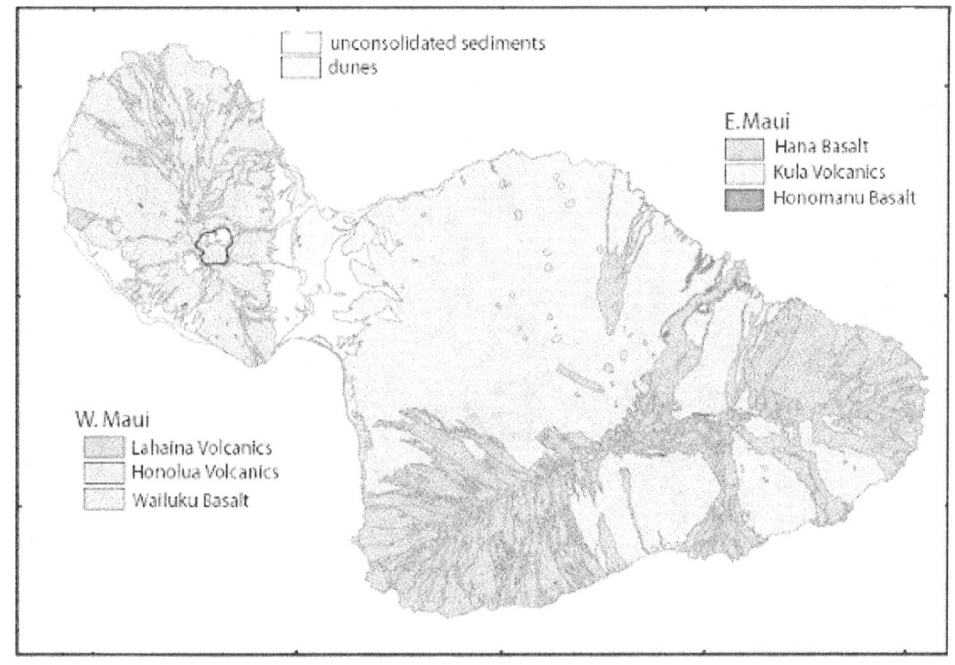

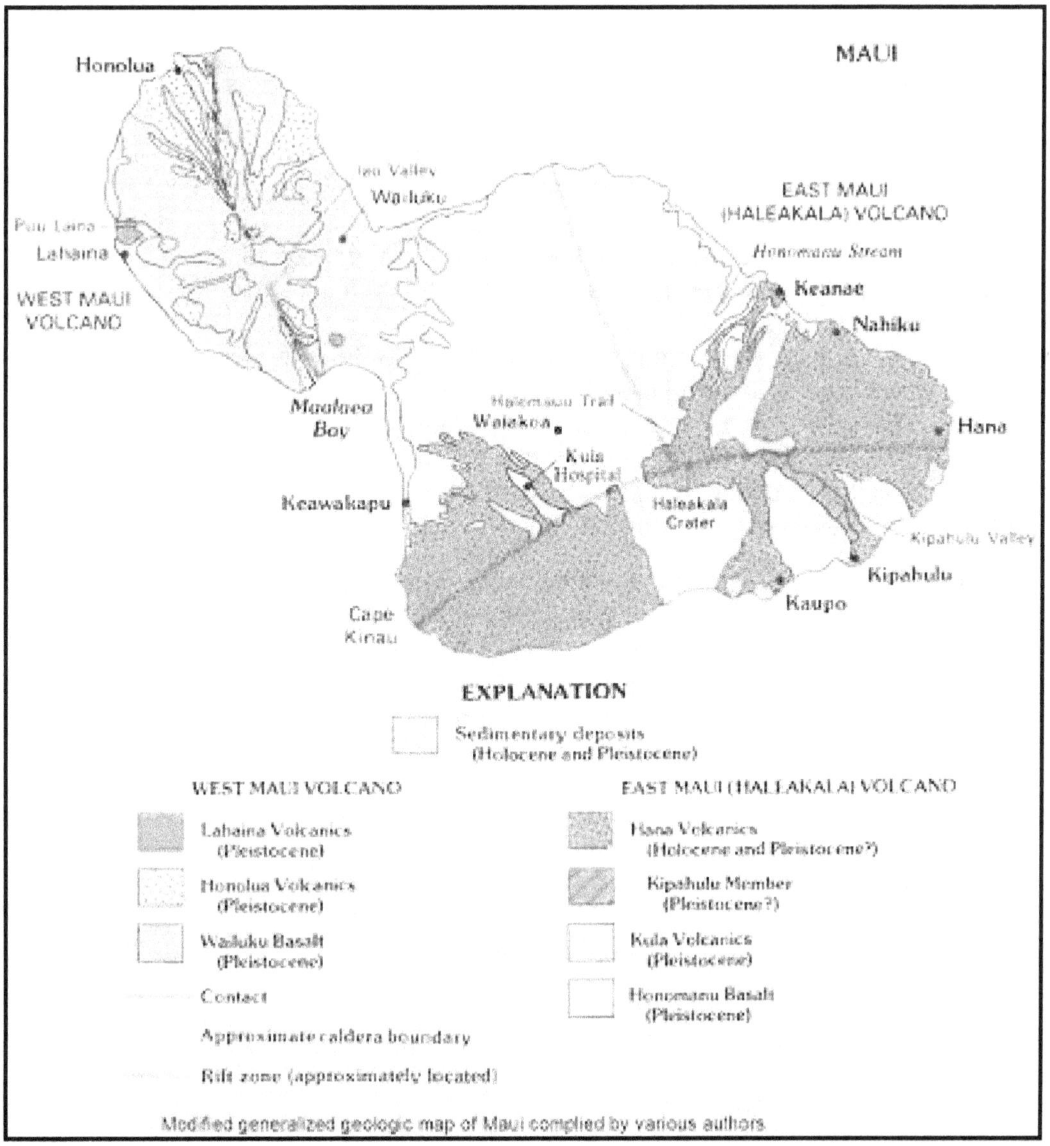

HALEAKALA ERUPTION HAZARDS

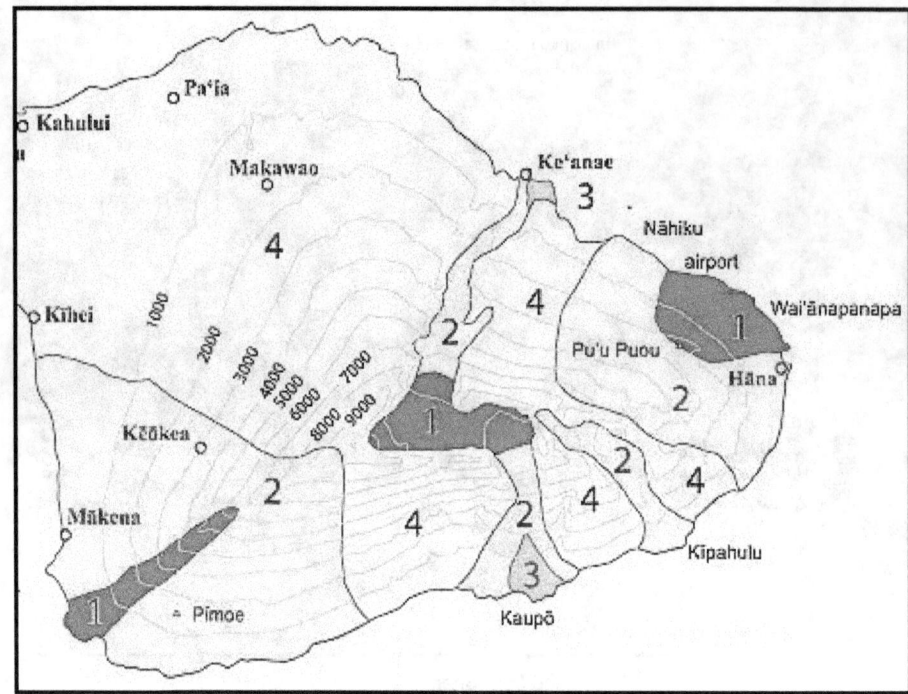

courtesy of the United States Geological Survey

Preliminary estimates of lava-flow hazard zones on Maui made in 1983 by the U.S.G.S.

- High hazard ratings were based on a zone having erupted three times in approximately the last 900 years.
- The lower the rating number the higher the volcanic eruption risk.
- However, the these frequency of eruptions *are only one criteria* on which these hazards are based.
- Other important eruption criterion include the lava flow coverage rate.
- Using the preliminary dates for Haleakala flows, only 8.7 square miles of lava has been emplaced in the last 900 years.
- However, this area of lava coverage is many times smaller than the area covered by the volcanoes on the Big Island of Hawaii.
- For the last 30,000 years, volcanism occurred on Haleakala's southwest and east rift zones.
- Maui's most recent lava flow, 1480-1600, is shown in red on the southwest flank of Haleakala.
- *However, websites say the last eruption occurred in 1790, which the USGS say is in error.*
- The oldest *exposed* lava flow on Maui dates at 1.1 million years ago.

ZONE 1: The most likely site of the next eruption will be along the Southwest Rift Zone from La Perouse Bay along the length of the Haleakala "Crater".
- While the East Rift Zone is generally less active, an area of vents upslope of Waianapanapa and the Hana airport are also rated as being in Zone 1 because each of these areas has had at least five eruptions in the past 1,500 years.
- However, no eruptions have occurred in the last 450 years.

ZONE 2: Encompasses the north and south flanks of Haleakala's Southwest and East Rift Zones.
- This zone includes land that was covered by lava at least once in the past 13,000 years.
- *Keanae Valley and the Kaupo area* are also in Zone 2 because they are down slope from lava flows that might erupt within the Haleakala "Crater".

ZONES 3 and 4: This area encompasses most of East Maui.
- Because Haleakala's "Crater" can block rift-zone lava flows from all but a few paths down slope, this area has essentially has no hazardous lava flow scenarios.
- Also, new vents erupt infrequently within Zone 4.

EARTHQUAKES ON MAUI
(Source: http://hvo.wr.usgs.gov/volcanowatch/archive/1996/96_11_27.html)

- Historic earthquakes have occurred on Maui.
- The University of Hawaii seismologists estimate that Maui County can expect a 3 to 5 magnitude earthquake every 2 to 5 years, a magnitude 6 earthquake about once every 50 years, and magnitude 7 earthquake about once every 250 years.
- By comparison, the southern coast of the Big Island can expect a magnitude 7 earthquake every 30 to 40 years and a magnitude 8 earthquake every 120 to 190 years.

MAUI'S 1938 EARTHQUAKE
Source: http://hvo.wr.usgs.gov/volcanowatch/archive/1999/99_04_08.html

- On January 22, 1938, at about 10:03 p.m., a magnitude 6.8 earthquake struck the central part of the Hawaiian island chain. It was located about 12 mi northeast of Keanae Point in East Maui.
- It was a submarine earthquake that originated at a depth of 12 miles. Thus. it was a shallow focus earthquake.
- The north side of Maui sustained most of the damage.
- But, landslides blocked the road to Hana and communications were completely severed.
- Ranches in southeastern Maui also had heavy damage, as water tanks and stone walls collapsed.
- In central and west Maui, from Kahului to Lahaina, concrete buildings cracked.
- In Kahului, the fire station tower shifted 0.5 inches (13 mm).
- Fortunately, no lives were lost and injuries were few.
- Also, no tsunami accompanied this earthquake. *Note:* To trigger a tsunami, the earthquake must have a magnitude of 8 or more. **Or**, a smaller earthquake can cause a tsunami if it triggers a large submarine landslide.
- The 1938 earthquake was one of the few Hawaii earthquakes felt throughout the islands.
- Up to that time, Kauai residents said it was the severest shock in memory.
- The 1938 Maui earthquake was *not related to volcanism*, as is common on the Big Island.
- Rather, it was a tectonically produced earthquake. This type of earthquake results from loading and bending of the Earth's crust by the volcanic mass of each island.
- These earthquakes diminish in frequency as each island moves off the Hawaiian hotspot and away from the zone of flexure associated with the larger islands.

MAUI'S 2016 EARTHQUAKE

- Early Monday morning on March 28, 2016, a magnitude 3.6 earthquake struck beneath the isthmus connecting West and East Maui.
- Then, four days later, late on the night of Thursday March 31, 2016, a magnitude 4.2 earthquake, centered roughly 55 miles east of Hana, struck.
- This earthquake was felt across the state.
- Unfortunately, seismologists state that Maui can expect more future earthquakes.

TSUNAMI HAZARDS ON MAUI

- Tsunamis are long wave length waves that travel across an ocean basin at speeds of 400 miles per hour or more.
- Tsunamis are often called "tidal waves", but they have *nothing* to do with the tides.
- Tsunamis form when the water in an ocean basin is disturbed by an earthquake, a submarine landslide, a volcanic eruption, or by a meteoric impact.
- These waves come in groups, and the first one is *not necessarily* the largest one.
- The famous April 1,1946 tsunami that struck Hilo, the highest tsunami wave was 55 feet high.
- Unfortunately, regarding the map on the next page, the publisher of this book do not allow pages to be printed in landscape view.
- Consequently, this map cannot be enlarged to allow the names of the towns to be legible.
- But, low lying coastlines susceptible to flooding can be seen in the large "rectangular" areas.
- Thus, if there is a tsunami alert, avoid driving in the red areas, and the diamond shaped areas, as they too are susceptible to localized flooding. **Continued on the next page.**

- For Maui tsunami evacuation zone maps go to: https://www.mauicounty.gov/261/Tsunami-Evacuation-Maps

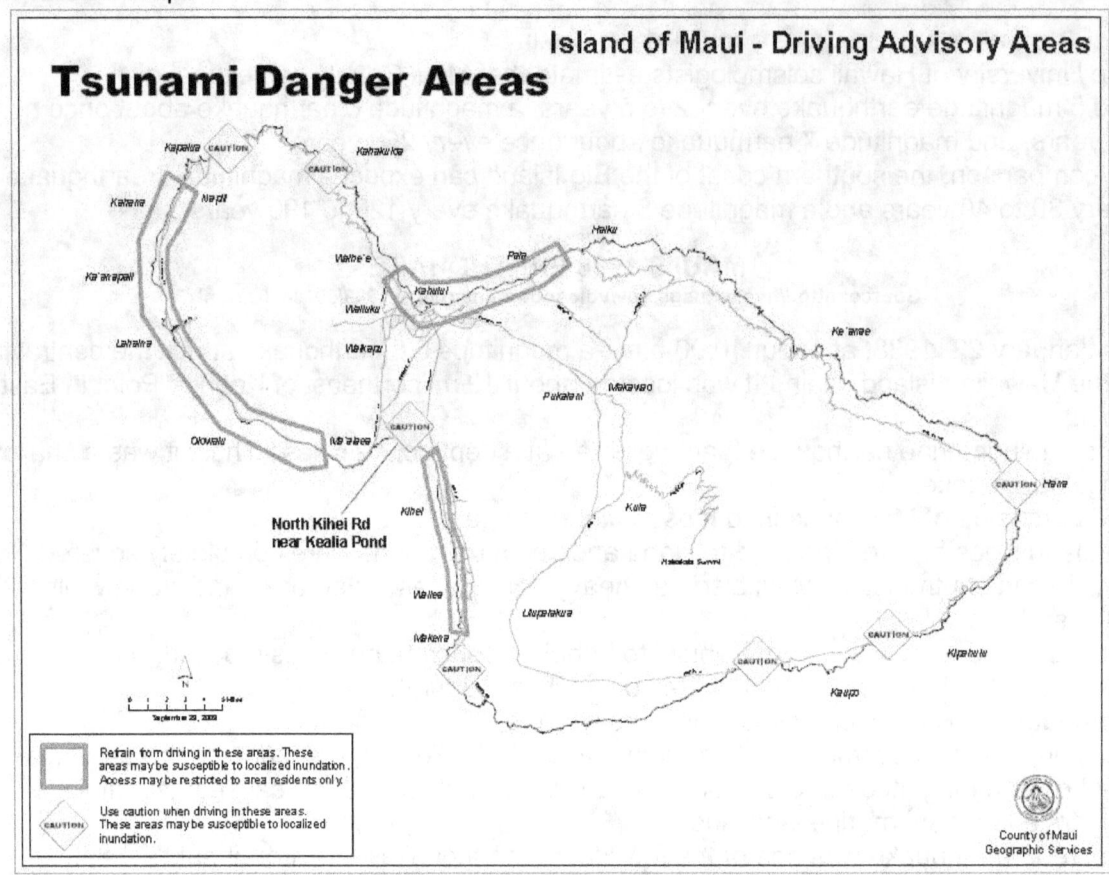

Courtesy: County Maui Geographic Services

THE PLEISTOCENE GLACIATION

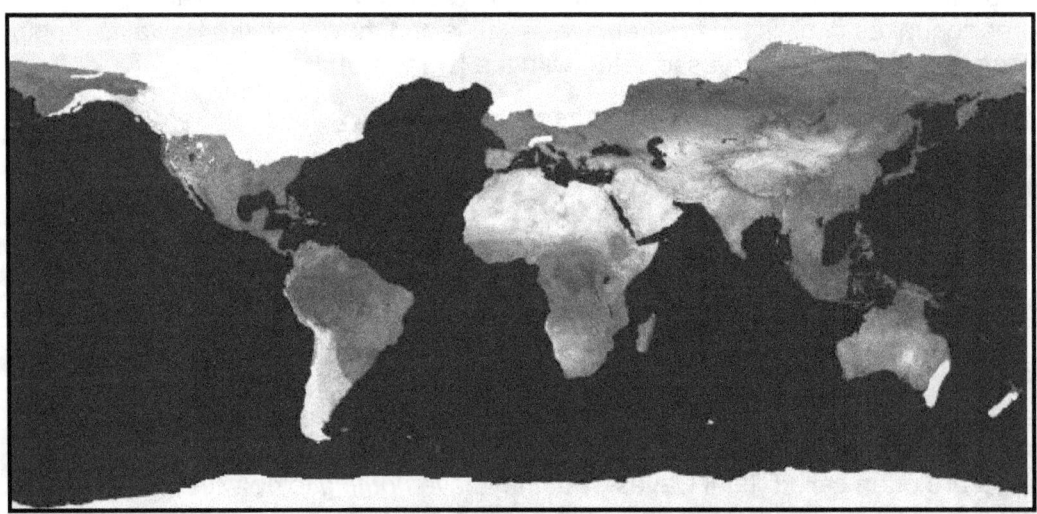

Modified with deep appreciation on an unknown source

- The above map shows the extent of the Pleistocene continental ice sheets.
- The Pleistocene glaciation occurred 2.6 million years ago until about 12,000 to 10,000 years ago.
- While there were other periods of glaciation during geologic time, the Pleistocene glaciation was the last glacial period in the earth's history.
- The Pleistocene glaciation was primarily in the northern hemisphere, but valley glaciers did form in the valleys of high mountains in the southern hemisphere. *See the above map.*

PLEISTOCENE GLACIERS ON HALEAKALA

- The Hawaiian Islands contain the only summits in the Pacific Basin that retain evidence of multiple Pleistocene glaciations.
- For glaciers to form on a Hawaiian volcano, it's summit had to be higher than the snowline.
- Because the *original* height of Haleakala was at least 1200 feet higher than its current elevation of 10,023, its summit had as many as 10 glacial advances.
- Snow first appeared on Haleakala's summit about 800,000 years ago.
- Then, the ice caps grew and shrank for the next 400,000 years (Thornberry-Ehrlich, 2011).
- But, after island subsidence, erosion, and reduced volcanic activity, the summit became lower than the snowline and glaciation ceased.
- However, during this period of glaciation, evidence of glaciation in the summit basin's has been obscured by young lava flows and cinder cones.
- In addition to summit glaciation, the Keanae Valley on the north side of the summit basin, and the Kipahulu Valley on the south side of the summit basin may have been shaped by glacial ice. Also, glacial triggered floods likely eroded the lower parts of these canyons (Thornberry-Ehrlich, 2011).

OPTIONAL THE CHEMICAL COMPOSITION OF HAWAIIAN LAVA

- There are two major types of basalts: *tholeiitic* basalts and *alkali olivine* basalts.
- *Alkalic olivine basalts* contain olivine and more sodium and/or potassium than the *tholeiitic* basalts.
- *Tholeiitic* basalts are richer in silica and iron and poorer in aluminum than alkalic basalt. They are rich in plagioclase feldspar and clinopyroxene with little or no olivine.

OPTIONAL OCCURRENCE OF THESE TYPES OF LAVA

- *Alkalic basalts* are the lavas extruded in the capping stage of Hawaiian volcanoes.
- In contrast, *tholeiitic basalts* are extruded in the shield-building stage. Thus, they make up the bulk of the Hawaiian volcanoes.
- Because *tholeiitic basalts* built the Hawaiian Islands and other mid-oceanic landmasses, they are more abundant on Earth than *alkali olivine* basalt.

ELEVATION MAP OF MAUI (Modified from hawaii-guide.com/maui/maui-map)

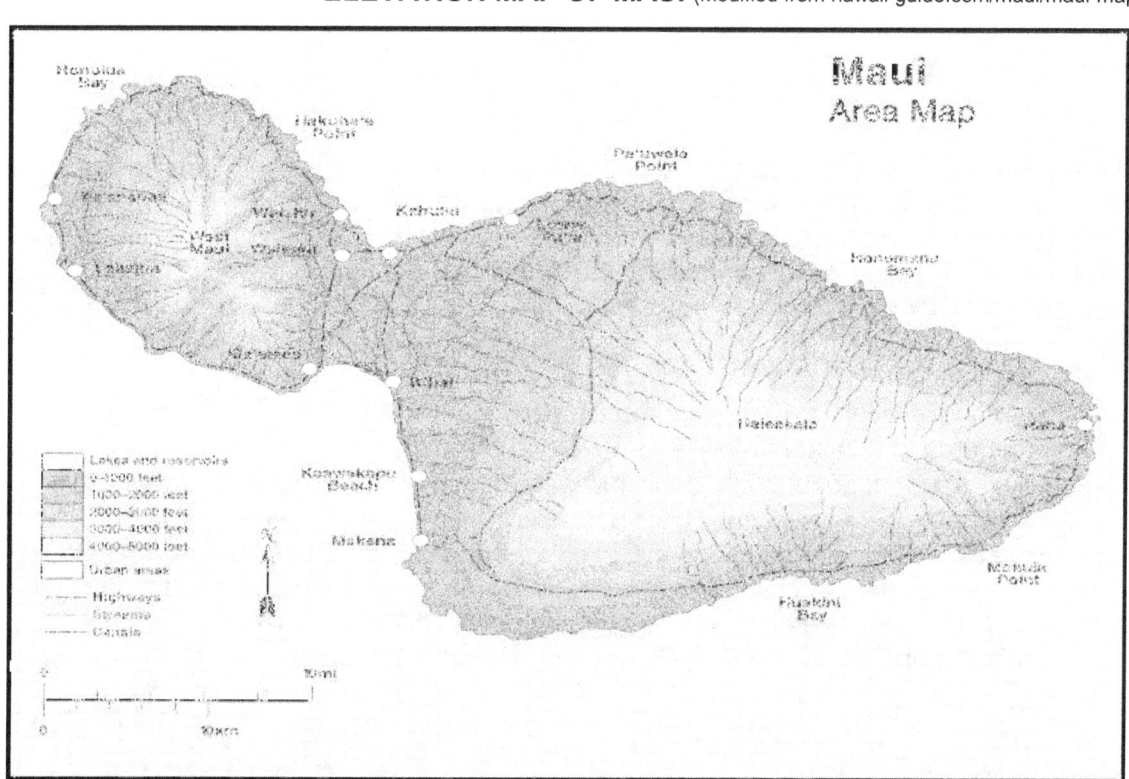

GEOLOGY OF THE HALEAKALA VOLCANO
"House of the Sun"

Haleakala (Hale-a-ka-la), literally, *House of the Sun*).
- The Island of Maui is composed of two volcanoes, the older and smaller West Maui Volcano, and the younger and much larger East Maui Volcano, most commonly known as Haleakala. *See the map below.*

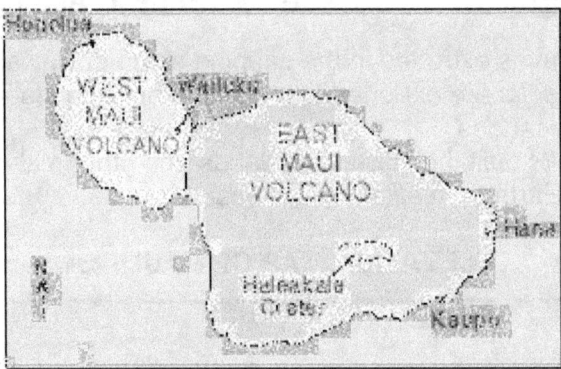
Courtesy of the USGS

- Haleakala is a large, broad shield volcano with gently sloping sides with small cinder cone volcanoes on both flanks. *See the first photo above.*
- Haleakala means: *the house of the sun.*
- A number of astronomical observatories are located on its summit (the small white dots at its summit in the above photo).
- Haleakala:
    - is located at: 20.718º N 156.254º W
    - has an elevation of 10,023 feet above sea level:
    - has area of 570 square miles, or 77% of the island.
    - has a volume: about 7,200 cubic miles, with 97% of this volume below sea level.
    - has a 33 mile diameter at sea level, and a much larger diameter on the sea floor.
- The name Haleakala is most commonly used for the *entire shield* of the East Maui Volcano.
- However, early Hawaiians used this name *only for the summit area*, where the demigod Maui snared the sun, and forced it to slow its journey across the sky. Go to:
https://www.skylinehawaii.com/blog/legends-of-demigod-maui

HALEAKALA'S DEVELOPMENT

- **Sequence of a Hawaiian Volcano's Eruptive History:** Hawaiian volcanoes *might* progress though the sequence of: *Pre-shield building, Shield building, Post shield building, and Rejuvenated stages. See page XXXX for a description of these stages.*
- **However,** Haleakala has not yet progressed through all of these stages. *See below.*
- **Oldest Dated Rocks:** are about 1.1 million years. These lavas were emplaced *near the end of its shield building stage.*
- **Time of Formation:** To build a volcano from the sea floor to the end of its shield-building stage is estimated to be about 0.6 million years. Thus, the Haleakala *(the East Maui volcano)* probably began its growth about 2.0 million years ago.
- **Previous studies** stated Haleakala was in a rejuvenated stage, but revised dating indicates *Haleakala is in the waning stages of a post-shield stage.*
- **Oldest on-Land Eruptions:** Haleakala's slopes are mantled by lava flows 700,000 to 150,000 years old. But, numerous small lava flows have occurred on its southwest and east rift zones in the past 30,000 years.
- **Youngest on Land Eruptions:** *The 1790 eruption was often stated as the most recent eruption. But, newer dating indicates the most recent eruption occurred sometime between 1480 and 1600.*

- Haleakala has two major rift zones (fractures in the earth's crust from which lava is extruded).
- One major rift zone extends from the summit to the east, and the other major rift zone extends from the summit to the southwest.
- A third, smaller, less active rift zone extends to the northwest. *See the map below:*

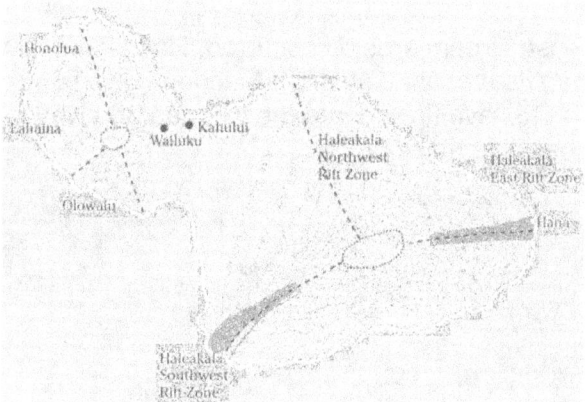

Courtesy of GeothermEX.Inc

- The Southwest and East Rift Zones form a curving arc from La Perouse Bay (on the southwest flank of Haleakala) through the Haleakala "Crater" to Hana on its east flank. *See the above map.*
- This arc continues east onto the seafloor as the Hana Ridge. *See the map below.*

Modified from USGS

- *Continued on the next page.*

Haleakala's Summit Basin, aka its "*Crater*"

- A large topographic basin, aka, Haleakala's "Crater" occupies its summit.
- This crater-like feature is about 20 miles in circumference.
- It is 7miles long, 2miles wide and nearly 2,625 feet deep
- This basin is *not volcanic in origin. Rather it is due to stream erosion.*
- Several hundred thousand years ago, streams in the Keanae and the Kaupo valleys, drained respectively, to the northeast and southeast eroding back into the East Maui Volcano.
- Then, about 120,000-150,000 years ago, the two valley heads were breached, forming the Koolau and Kaupo gaps and this "crater-like" depression. *See the illustration below.*
- The **Kaupo Gap** is the large, wide stream eroded opening at the south side (i.e., bottom) of Haleakala's summit.
- The **Koolau Gap** extends in a northward direction from the rim on the Haleakala basin.

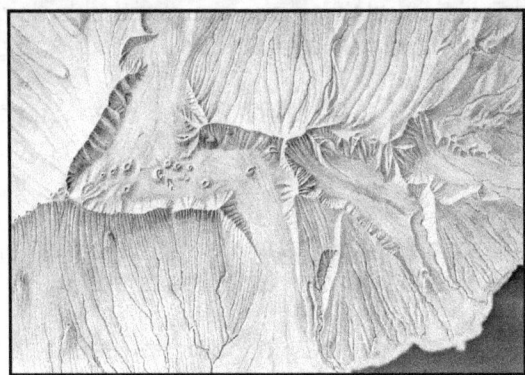

Modified from National Park Service

- Before this period of intense stream erosion occurred, Haleakala's summit may have been 2,000-3,000 ft. *higher than its present elevation* (Macdonald 1978).
- Later, from the volcano's rim, lava flowed into the Keanae and Kaupo valleys and into the sea.
- An image of the "crater/basin" is show below.

Courtesy of the National Park Service

- For a better view of this image, go to **http://npmaps.com/haleakala**

HALEAKALA'S ERUPTIVE HISTORY

- Haleakala is composed of ropy (pahoehoe) lower viscosity lava, blocky (a'a) higher viscosity lava flows, and thin pyroclastic deposits (ash and cinders).
- These volcanic products record a nearly complete sequence from the early shield building stage to late stage lava capping. See page 19.
- Haleakala had three major eruptive stages (from oldest to youngest): the Honomanu lavas, the Kula Volcanics, and the Hana Volcanics.

- The **Honomanu shield-building phase:** ropy and blocky basalt, with few surface exposures, found mainly in deep canyons. Extrusion of this lava ended approximately 750,000 years ago, but most lava ages date from 1.1 to 0.97 million years.

- The **Kula post-shield building phase:** composed of *Hawaiite* basaltic lavas. In contrast to the Honolua Volcanics, the Kula Volcanics have more ash and cinders produced by explosive eruptions that formed many large cinder cones. Kula lava are exposed in the walls of the Haleakala "crater" and as surface lava flows, especially in the road cuts seen as you drive to Haleakala's summit. Kula Volcanics range in age from 930,000 to 150,000 years (USGS).

- The **Hana post-shield building phase:** are lavas extruded in renewed rift zone volcanism. These lavas are similar in composition to the Kula Volcanics, but they are thinner and were more fluid and explosive. Hana eruptions formed large cinder cones. These eruptions began about 120,000 years ago.

- Haleakala *is not in the classically-defined rejuvenated stage* because there is no evidence of a lengthy gap between the youngest Kula Volcanics at 150,000 years and the oldest Hana Volcanics at 120,000 years. Thus, *Haleakala is in the waning stages of the post-shield volcanic stage, and is considered to be a potentially active volcano. See the distribution of these units on the map on the below.*

GEOLOGIC MAP OF HALEAKALA'S LAVA FLOWS

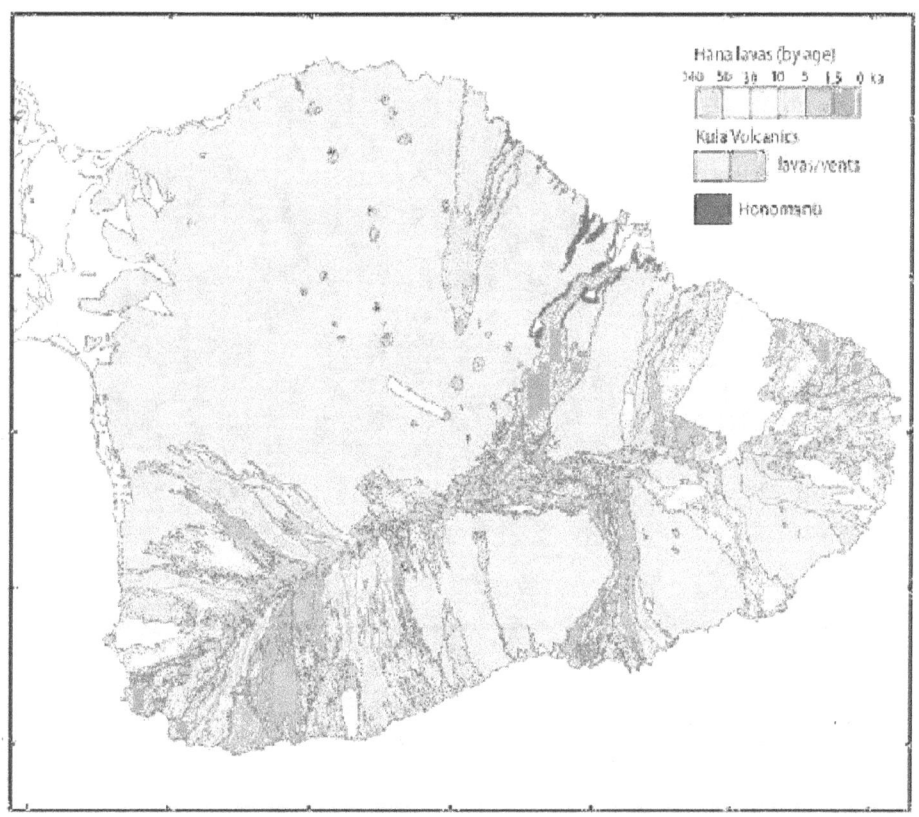

Modified from Sinton 2006

- *Continued on the next page.*

- As previously stated, lava from Haleakala has been mainly erupted from two major rift zones.
- The previous geologic map shows Haleakala's Northeast Rift Zone *mainly* in greenish colors, and the Southwest Rift Zone in colors of purple, red, green and yellow.
- These rift zones can also be seen the digital elevation image below:

Courtesy of USGS

MAP OF LAVA FLOWS IN THE SUMMIT BASIN

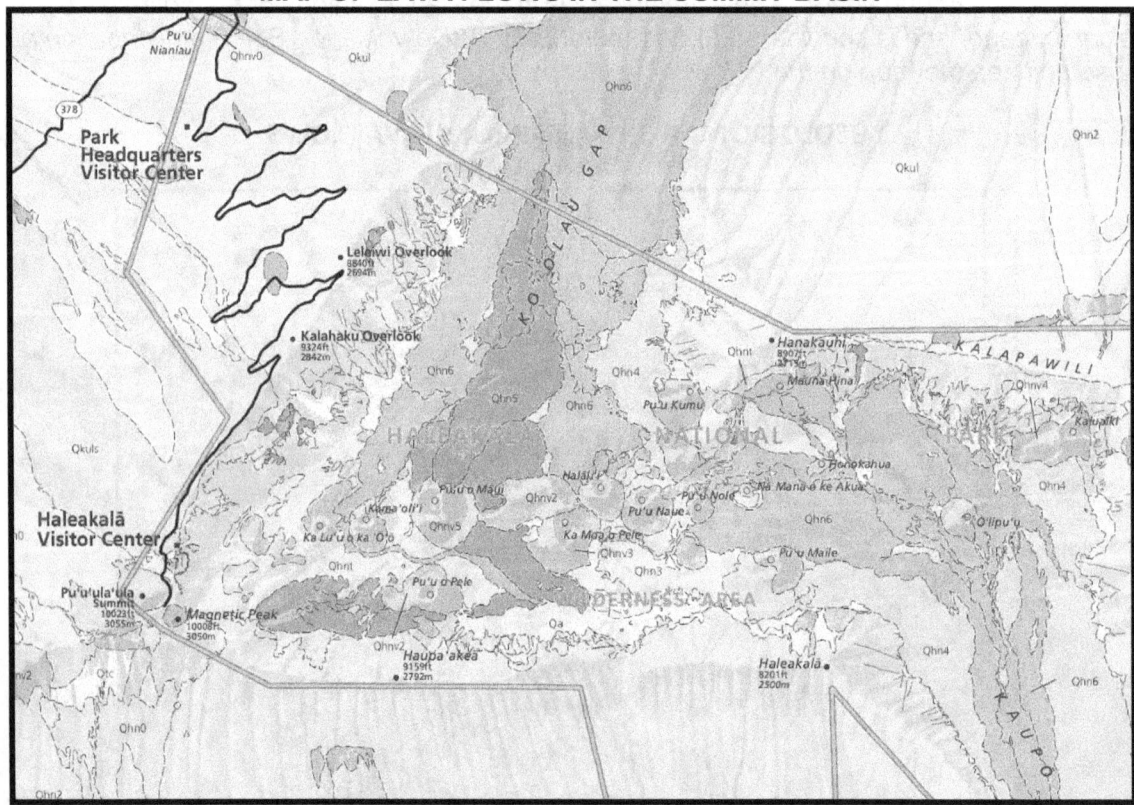

Courtesy of the National Park Service

- The publisher of this book will not allow any pages in landscape view.
- **Thus, for a readable** view of this map, go to **http://npmaps.com/haleakala**
- An enlarged view of this map would show fourteen multicolored cinder cones in the summit basin of Haleakala.
- Cinder cones form when gas is trapped in lava during an eruption and forces the lava to eject as a fountain. • *Continued on the next page.*

- The resulting hot lava falls as sticky cinders all around the base of the fountain to from a cinder cone.
- Cinder cones are small volcanoes generally no higher than 1,200 feet. *Photo to the right a Haleakala summit cinder cone.*
- Cinder cones usually erupt lava flows, either through a breach on one side of the crater, from a vent located on its summit, or from its flank.
- Cinder cones are commonly found on the flanks of shield volcanoes and also in a volcano's caldera (a large summit opening larger than a crater).
- Today, the summit of Haleakala stands at 10,023 feet, but it once may have reached 15,000 feet above sea level.
- But stream erosion, combined with the weight of Haleakala resting on the earth's crust, has caused the crust to subside.
- Thus, when measured from the seafloor, Haleakala is 29,703 feet above the sea floor.
- Consequently, Haleakala is taller than Mount Everest by about 675 feet.
- Interestingly, satellite measurements recorded that Haleakala rose 0.23 ft. (.07m.) per year between 1981 and 1992.
- Geologists theorized this increase in elevation could be due either to the movement of magma deep under the sea floor, **or** the modern day eruptions on Hawaii Island added mass to the island causing the Big Island to subside, thus forcing Maui to rise up.

HALEAKALA ERUPTION HAZARDS
(Repeated for the sake of completeness)

Modified map of preliminary estimates of Maui lava-flow hazard zones made in 1983

- Because Haleakala has had historic eruptions, it is considered to be an active volcano.
- **Hazard rating:** The lower the number on the above map, the higher the volcanic hazard.

ZONE 1: Future eruptions are most likely to occur on the Southwest Rift Zone from La Perouse Bay to the Haleakala Crater.
- Also, while the *East Rift Zone* is generally less active, an area of vents upslope of Waianapanapa and the Hana airport are also rated as being in Zone 1.
- Each of the Zone 1 areas has had at least five eruptions in the past 1,500 years.
- However, no eruptions have occurred in this zone in the last 450 years.

ZONE 2: The north and south flanks of Haleakala's Southwest and East Rift Zones, includes land covered by lava at least once in the past 13,000 years. *Keanae Valley and the Kaupo area are in Zone 2 because they are down slope of lava flows that might erupt within Haleakala's Crater.*

ZONES 3 and 4: Most of East Maui. Except for a few paths down slope, Haleakala's crater blocks rift-zone lava flows. Thus, this area has essentially no hazardous lava flow scenarios. Also, new vents erupt infrequently within zone 4.

SUMMARY OF HALEAKALA'S WEATHER

Annual temperatures:
- As you drive to Haleakala's summit the temperature decreases about 3° F. per 1000 feet of elevation. This is why you often need to bring a jacket or sweatshirt when you drive to the summit. Need another reason, read below.
- The average *annual* temperature at the park headquarters located at an elevation of 7000 feet in elevation is 53 degrees F. But, the *annual* temperature at the 10, 023 foot summit is even colder.
- But, at the park headquarters, on any given day it can be warm because the highest recorded temperature was 80 degrees F. (October, 1973).
- Unfortunately in April, 1969, January 1982, January 1990 the lowest recorded temperature was 30 degrees F.
- What about wind at the summit? In December 1990, a wind indicator near the summit broke in at 128 mph.

Average annual rainfall:
- At the park headquarters at 7000 feet, the average annual rainfall is 53 inches.
- But, at the 10,023 foot summit, it decreases to an average annual rainfall of 40 inches.
- However, in the Kipahulu of the Haleakala National Park, the average annual rainfall is 187 inches.
- However, the maximum amount of rainfall in a 24 hour period occurred in January 1980 when 18.5 inches of rain was recorded.

LOCATION OF MAUI TOWNS IN RELATION TO THE ABOVE RAINFALL

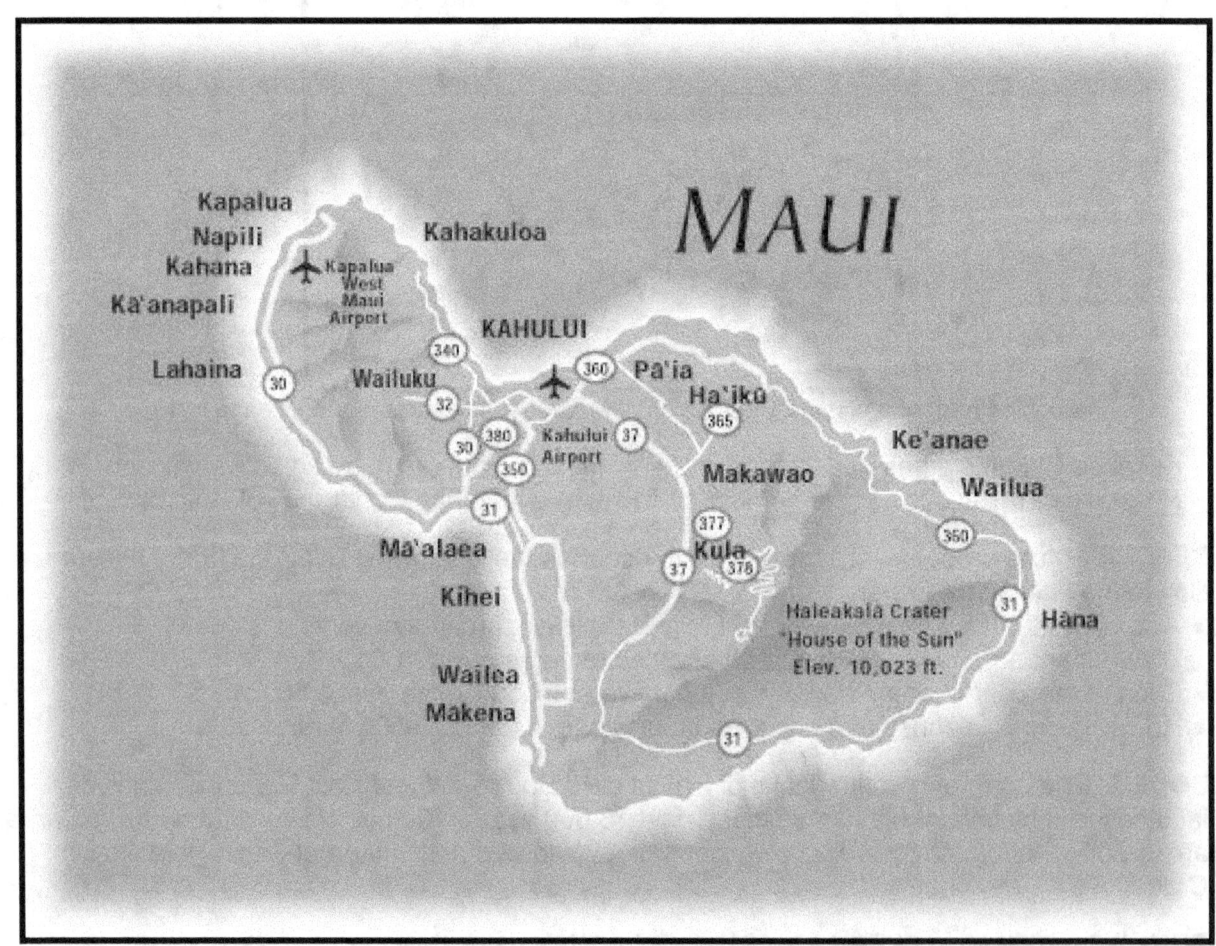

Courtesy of Hawaii City.com

HALEAKALA DURING THE PLEISTOCENE GLACIATION
(Repeated for the sake of completeness)

- The Pleistocene Glaciation began about 2-3 million years ago, and except for high mountain ranges in the Southern Hemisphere, this period of glaciation was mainly confined to the Northern Hemisphere. *See the map below.*

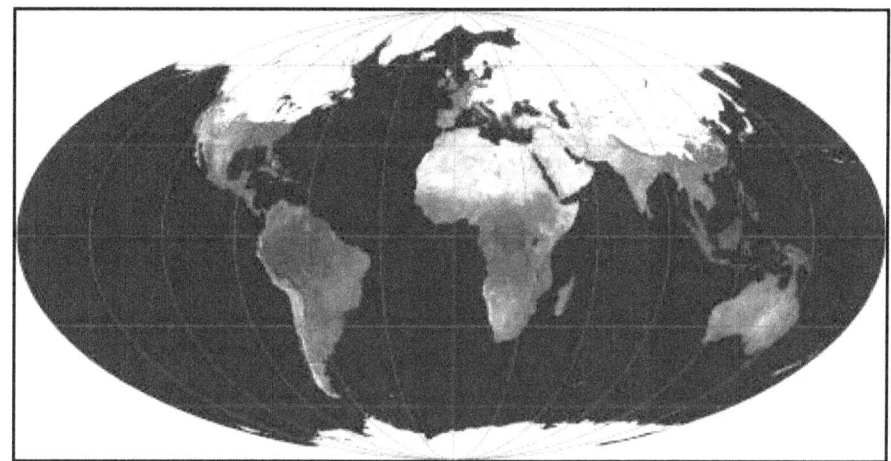

Map courtesy of NASA.gov.

- However, in the Pacific region, Pleistocene glaciation is only found in Hawaii, because the volcanic summits were higher than the snowline (the line above which snow can accumulate).
- Thus, because Haleakala's *original* summit had an elevation of at least 13,000 feet, it is likely Haleakala underwent Pleistocene glaciations. However, any glacial deposits on its summit, have since been buried by younger lava flows.
- Snow probably first appeared on Haleakala's summit about 800,000 years ago.
- The resulting ice caps expanded and retreated throughout the next 400,000 years. But then, glaciation ceased when the volcano's summit was lowered below the snowline by 1) the sinking of the island (due to its weight), 2) stream erosion, and 3) reduced eruptive activity (Porter 2005).
- Compared with the glaciations on Mauna Kea on the Big Island, some geologist believe Haleakala may have experienced up to 10 glacial advances.
- These glaciations may have eroded the Haleakala crater and the upper reaches of the two large canyons emanating from it (Thornberry-Ehrlich, T., 2011).
- In addition, multiple glacial generated floods probably eroded the lower reaches of these canyons (Thornberry-Ehrlich, T. 2011).

GEOLOGY AND EARLY HAWAIIAN SETTLEMENTS ON HALEAKALA

- The location of villages and farming on the slopes of Haleakala was determined by:
  1) the types of lava present.
  2) the soil.
  and 3) the climate.

- These factors **restricted farming** and the development of the *early* Hawaiian villages, to areas **between** arid zones **at low** elevations and a zone **at higher** elevations (where the soils were depleted in nutrients). Thus, farming of sweet potatoes was only possible **between** these to areas.

- Thus, on Haleakala's **southern arid** side:
  - in areas of **older Kula lavas** overlain by **more weathered volcanic ash,** nutrient-rich, fine-grained soils were present, and hence, large farming villages were built.
  - on **younger, less weathered Hana lava flows,** home sites and other structures were built.
- For details see: **www.pnas.org/content/101/26/9936**   • *Continued on the next page.*

GEOLOGY RELATED TO HAWAIIAN SETTLEMENTS

- By 1200 AD, Hawaiians established villages *in various environmental settings* on the slopes of Haleakala.
- Favorable geologic settings and adequate rainfall were found at elevations of 1,300 to 1,968 feet above sea level.
- Coincidently, this is approximately the elevation of the present day Upcountry Kula farming area, which has an elevation of 1800-3700 feet above sea level.
- Thus, these *modern day* Maui farmers are continuing a Hawaiian tradition.
- However, the Hawaiians were not confined to the lowland areas, because in 1400-1600 AD, they also had temporary campsites in alpine and sub-alpine elevations 1.2 to 2 miles above sea level.
- These temporary high altitude excursions were probably undertaken to collect birds and a *high quality of unweathered basaltic rocks used to make stone tools*.

GEOLOGY'S RELATIONSHIP TO THE EARLY HAWAIIAN SETTLEMENTS ON HALEAKALA IN THE KAHIKINUI REGION

- Kahikinui is located on Haleakala's *southern, arid flank* between the Ulupalakua Ranch and Kaupo on the Piilani Highway (Route 31). *See the map to the lower right.*
- The aridity of this area, due to a rain shadow, is evident *in the image to the lower left.*

Courtesy of the National Park Service

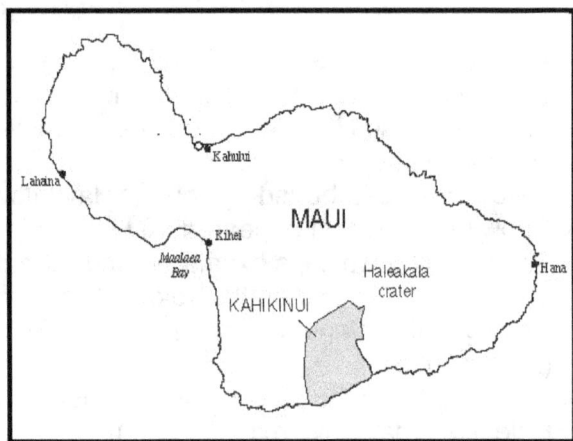
Courtesy of USGS

- Maui's rain shadow forms as storm clouds approach Maui from the east.
- As these clouds move up Haleakala and the West Maui Volcano, they lose most of their moisture.
- Hence, the *eastern* side of Maui has abundant green, vegetation.
- But, as the air descends down the *western* side of these volcanoes, these moisture depleted clouds begin to warm; and hence, they can better hold what little moisture they contain.
- Thus, the western side of Maui is in a rain shadow created by these two volcanoes.
- Consequence, Maui's western side is an area of dry grass and other desert-like vegetation.
- The formation of a rain shadow is *illustrated in the diagram below:*

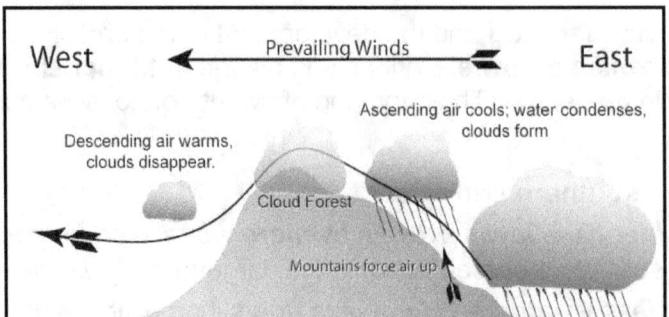
Modified from Marietta College.edu

- *Continued on the next page.*

- The Kahikinui area's permanent farming villages were established by 1400 AD
- However, these villages were on marginal farm land, and the population was sustained by growing sweet potatoes, dry land taro, and yams.
- But, by 1700-1800 AD, the population density of 43 to 57 persons per square kilometer peaked.
- As these early Hawaiians left this area for more favorable regions, they left behind more than 3000 archaeological sites (Kirch P.V., 2004).
- One of these archeological sites can be seen driving from the Ulupalakua Ranch to Kaupo.
- Currently, the Department of Hawaiian Home Lands is leasing plots of land to people who are a least 50% Hawaiian and want to live here in the old Hawaiian way.
- But, a lack of water and other facilities has slowed this development.
- If you drive from the Ulupalakua Ranch to Kaupo, you will pass through this area, and you can see that living here takes a particular type of person.

A MILE BY MILE GEOLOGICAL GUIDE ALONG THE ROAD TO HALEAKALA'S SUMMIT IS GIVEN IN THE BOOK BELOW. THIS BOOK IS AVAILABLE ON AMAZON.

GEOLOGICAL GUIDE TO HALEAKALA NATIONAL PARK

BY

RICHARD ROBINSON

PREPARING TO DRIVE TO THE *SUMMIT* OF HALEAKALA NATIONAL PARK

- Start with a full tank of gas. *Gas is **not** available in the park.*
- Leave early!
- It will take several hours to reach the summit view site. See chart below.
- Leaving late in the day could mean that the summit is engulfed in cloud cover.
- No food or drinks are available in the park. Bring food and water, juice, or cola with you.
- Bring sunscreen
- Bring warm clothing.
- Day time temperatures can be significantly *cooler* than the temperature at your hotel.
- The weather could be warm, cold, or rainy. Bring clothing for all possible weather.
- Restrooms are available.
- *Very approximate* driving times from west shore resort areas to Kahului:
    - From Hana: 120 minutes
    - From Kaanapali: 50 minutes
    - From Kihei: 15 minutes
    - From Lahaina: 40 minutes
- These estimates do **not** include traffic delays. So, allow more time than these rough estimates!
- *Time to drive from Kahului to Haleakala's summit:* One hour and 30 minutes, depending on the traffic,

ROUTE FOR KAHULUI TO THE *SUMMIT* HALEAKALA NATIONAL PARK

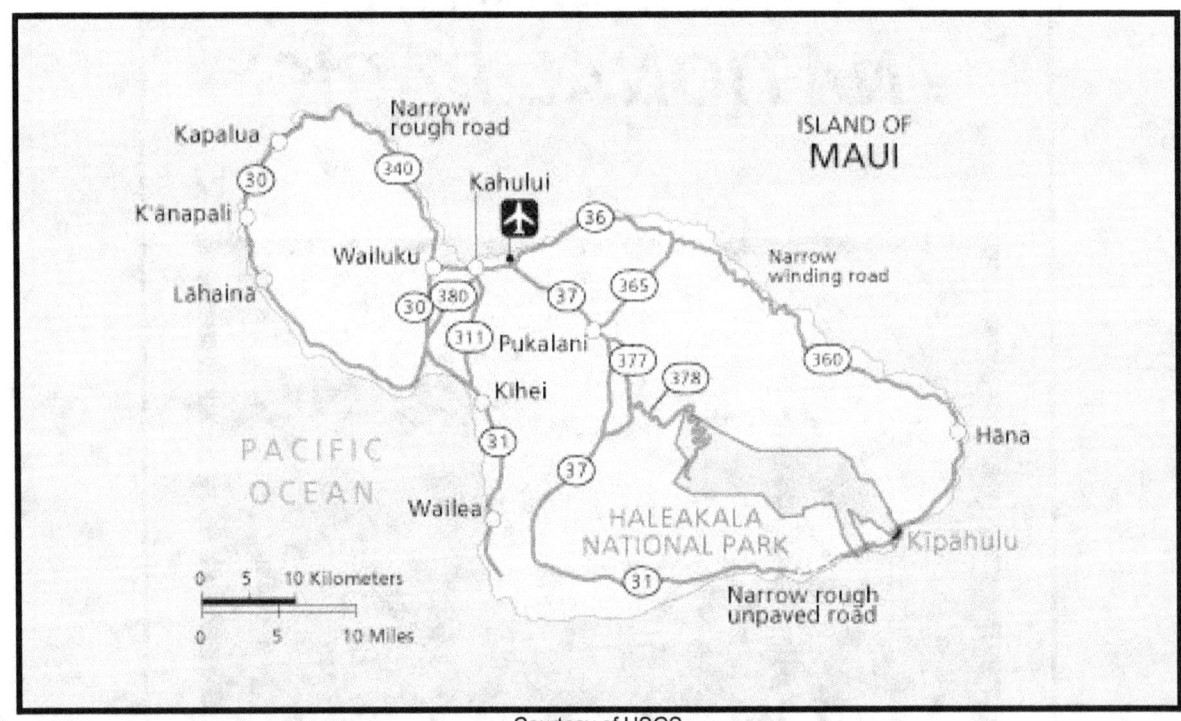

Courtesy of USGS

- Leave Kahului by going south on Route 36.
- At the intersection of Route 36 and Route 37, turn right and proceed on Route 37.
- At the intersection of Route 37 and Route 365, proceed on Route 37.
- At the intersection of Route 37 and Route 377, turn left and proceed on Route 377.
- At the intersection of Route 377 and Route 378, turn left and proceed to Haleakala's summit on Route 378.

....

VIEWING SUNRISE AND SUNSET ON HALEAKALA

Sunrise and Sunset on Haleakala:
- One of Maui's most exciting experiences is seeing the Haleakala sunrise or sunset.
- See the video link: www.youtube.com/watch?v=nBrzE5M-rm0
- To see this inspiring site takes some planning.

Make A Reservation!!
- The National Park Service requires a reservation for visitors in personal and rental vehicles to enter the Summit District from 3:00 am to 7:00 am to view the sunrise.
- The one-day $1.50 sunrise reservation can be made up to 60 days in advance.
- For sunrise reservations go to: **https://www.recreation.gov/ticket/facility/253731**
- **Note: The reservation fee is not part of the entrance fee.**
- **Entrance fees** will be collected upon entry to the park.
- Visitors with national park passes, please have your pass and ID ready to present at the gate.
- **Reservations are only for sunrise** and can only be used on the day that was reserved.
- Summit conditions cannot be predicted to say if the sunrise is going to be clear or cloudy.
- **Weather** is unpredictable, but it is often windy and wet.
- There are NO weather predictions available for sunrise. Sunset is slightly more foreseeable.
- Temperatures immediately before dawn and immediately after dusk are regularly below freezing.
- **There will be no refunds or exchanging reservations for a different day.**
- Note: The visitor centers are not open at these times.
- The road up and down the mountain does not have streetlights.
- It takes approximately 1.5 hours to drive between Kahului and the summit.
- Parking is restricted to designated lots only.
- Parking lots will be closed when full, especially at sunrise.
- **NOTE: No reservations for SUNSETS are required at this time (2019).**
- Sunset at Haleakala, with flaming clouds, is described as breathtaking the sunrise.
- Parking at the Haleakala Summit is limited. This reservation ensures that you will have a parking space at one of the four sunrise viewing locations at the summit.
- First-come, first-served. **Sunrise summit reservations are not available at the entrance station** or at the park.
- Reservations are now available, **online only**, up to 60 days in advance of your sunrise visit.
- You can also call the reservation line at 1-877-444-6777 to make a reservation over the phone.
- Calling the park directly, or visiting in-person, will not result in a reservation since staff at Haleakala National Park are unable to make reservations for you.
- A small number of last-minute tickets are released **online** two days beforehand at 4:00 PM
- The website will show tickets as sold out until 4:00 PM.
- As of February 2018, **Viewing sunset did not require reservations.**
- Depending where you are located on Maui, it can up to two hours to drive the Haleakala Highway (HI 37) and then up the winding Haleakala Crater Road leading to the summit of Haleakala.
- For a good viewing location, try to arrive at the Haleakala Visitor Center no later than a half hour before the sun rises.
- **In the summer the sun rises as early as 5:38 AM, in the winter as late as 6:55 AM.**
- The color of the sky and clouds before daybreak are stunning.
- Bring warm clothes! It gets very cold at the summit.
- The temperature drops 3° for every 1000 feet of elevation, so at the Haleakala Visitor Center's 9,740-foot elevation (where most people watch the sunrise), it's about 30° colder than at sea level.
- Be sure to bring pants, shoes, layers of clothing, and blankets. Temperatures are often in the 40-degree range. Don't arrive wearing shorts.

- Restroom facilities are available, but no food or gas sold within the park.
- *Continued on the next page.*
- Thus, start with a full tank of gas, and bring some food and drink.
- And most importantly don't forget your camera.
- Stay beyond sunrise!
- One of the most common mistakes visitors make is to leave immediately after the sun rises.
- If you stay ten to twenty minutes later you'll see an incredible show as the colors of dawn stretch across the landscape of the Haleakala National Park.
- **Sunset**
- Explore the park before the sun goes down.
- Sunset at Haleakala, with flaming clouds, is described as breathtaking the sunrise.
- Again bring some food and drinks.

,**Average Sunrise and Sunset Times**

- Summer sunrise average time is 5:45-6 a.m. HST and Winter sunrise average time is 7 a.m. HST,
- Sunrise times by day can be at: https://sunrisesunset.willyweather.com/hi/maui/haleakala-national-park.
- For more information: (808) 572-4400; or go to http://www.nps.gov/hale to make a reservation!

Average Sunrise and Sunset Times

- **Summer** sunrise average time is 5:45-6 a.m. Hawaii Stand Time and **Winter** sunrise average time is 7 a.m. Hawaii Standard Time.
- Sunrise times by day can be at: https://sunrisesunset.willyweather.com/hi/maui/haleakala-national-park.
- For more information: (808) 572-4400; or go to http://www.nps.gov/hale to make a reservation

SUMMIT PRECAUTIONS

- At Haleakala's summit you will be at an elevation of 10,020 feet. At this elevation the air is thinner, so:
    - do not walk or run vigorously or you may suffer high altitude induced headaches.
    - the temperature will be cooler at 10.023 feet than at your sea level resort.
    - be prepared for the summit being warm, cold, or rainy. Bring clothing for all conditions.
    - you may need sunscreen. You can get sunburned even on a cloudy overcast day.
- Remember that no food, bottled water, or other drinks are sold at the summit or visitor center.

SUNSCREEN

- Beginning in 2011, the State of Hawaii has banded all sunscreen products containing oxybenzone and octinoxate.
- Reason: A 2015 study determined that oxybenzone "leaches the coral reefs of their nutrients and bleaches them white. In addition, it can also disrupt the development of fish and other wildlife".
- These chemicals are used in more than 3500 of the world's popular sunscreens.
- **However,** sunscreen manufactures say that, with in the United States, other ingredients effective for SPF 50 have limited availability.
- **Furthermore,** the Hawaiian Medical Association wants this issue studied more deeply because the evidence suggesting oxybenzone was the cause of coral bleaching was not *peer reviewed.*
- The Hawaiian Medical Association also feared that not wearing any sunscreen would increase cancer rates.
- This issue is left for you to decide.

GLOSSARY OF SELECTED VOLCANIC TERMS
(Modified from: USGS Photo Glossary of volcanic terms.htm)

Aa or blocky lava (pronounced "ah-ah"): a Hawaiian term for lava flows with a rough rubble-covered surface composed of broken lava blocks called clinkers. The irregular surface of a blocky flow is very difficult to walk over. Please use caution walking across a blocky lava.
- forms as a slower, cooler, and hence more viscous lava flow advances across a landscape.
- the forward movement of this slow moving lava flow causes the cooled upper crust to break into irregular shaped blocks that tumble down the front of this slow moving lava flow.
- often contain gas bubbles, called vesicles, produced by gases escaping from the lava when it was extruded onto the surface.
- can be seen on the south side of Haleakala.

Ash: consists of rock, mineral, and volcanic glass fragments smaller than 2 mm (0.1 inch) in diameter, but particles less than 0.025 mm (1/1,000th of an inch) in diameter are common.

Basalt: a hard, black volcanic rock that formed from a cooling lava. Some basaltic lava flows are very fluid, and they can quickly flow more than 12 miles from a vent. Basalts are erupted at temperatures between 1100 to 1250° C. Most of the ocean floor is made of basalt. The shield volcanoes of the Island of Hawaii are composed almost entirely of basalt.

Blocks: solid rock fragment greater than 64 mm (2.52 inches) in diameter ejected from a volcano during an explosive eruption. Blocks commonly consist of solidified pieces of old lava flows that were part of a volcano's cone.

Bomb: droplets of lava ejected from a volcano while they were partially molten.

Calderas: a large, usually circular, depression at the summit of a volcano. Calderas form when large volumes of magma are withdrawn into the subsurface or erupted from side vents. With the structural support for the overlying rock removed, the walls of the volcano slump inward to form this large summit opening.

Craters: circular summit depressions much smaller than that of a caldera.

Cinders: partly vesiculated (contains gas bubble holes) basaltic lava fragment ejected during an explosive volcanic eruption.

Cinder cones: steep, conical hills, generally less than 1500 feet high, composed of volcanic fragments that accumulated around a volcanic vent

Dikes: tabular rock bodies that cross cut the layering of adjacent rocks.

Ejecta: a general term for rock particles ejected into the air by a volcano.

Eruption clouds: vertical columns of tephra (volcanic fragments) and gases rising directly above a vent as an eruption column. May drift for thousands of miles downwind.

Faults: fractures in the Earth's crust where one side moves with respect to the other. A fault scarp is a cliff or steep slope that sometimes forms along the fault at the earth's surface.

Fissures: elongate fractures or cracks from which lava erupts.

Gas: Magma and lavas contain dissolved gases that are released into the atmosphere during a volcanic eruption. The most common gas released from a volcano is steam (H_2O), followed by CO_2 (carbon dioxide), SO_2 (sulfur dioxide), (HCl) hydrogen chloride, and other compounds.

Lava: molten rock erupted onto the Earth's surface.

Lava flows: masses of molten rock that poured onto the Earth's surface during an effusive volcanic eruption. • *Continued on the next page.*

Lava fountains: jets of lava sprayed into the air by the rapid formation and expansion of gas bubbles in the molten rock. Lava fountains typically range from 30 to 300 feet in height, but occasionally they reach more than 1600 feet high.

Lava tubes: natural conduits through which lava travels beneath the surface of a lava flow. Tubes form by the crusting over of lava channels and pahoehoe (ropy) flows. At the end of an eruption, the lava in the tube drains away to form hollow lava tubes.

Lava velocities: The fastest recorded Hawaiian lava flow was from the 1950 Mauna Loa southwest rift zone eruption. This flow advanced from its vent to Highway 11 at an average speed of 6 miles/hour. In contrast, a typical Big Island Puu Oo blocky **flow** moved less than 1/3 of a mile/hr. Lava can flow faster than 6 mi/hr if the lava is in a lava tube or channel; where the lava is confined and can stay hot. In 1984 Mauna Loa eruption, lava velocities within a lava channel were measured at nearly 35 mi/hr. During the first years of the Puu Oo eruptions, lavas **within lava tubes** had speeds up to 23 miles/hr.

Magma: molten rock beneath the Earth's surface. When magma erupts onto the surface, it is called lava.

Minerals: naturally occurring substances that are inorganic in composition, have an orderly internal arrangement of their atoms or ions, have a definite or nearly definite chemical composition, and have definite physical properties, such as density and hardness. In contrast, *Rocks* are composed of one or more minerals

Obsidian: a glassy *massive* igneous rock that forms when a lava flow cools so rapidly that minerals do not have time to crystallize and a natural glass forms. Obsidian is usually black in color due to the presence of iron oxide (magnetite, Fe_3O_4) within the glass. If the obsidian is highly oxidized, hematite Fe_2O_3 gives the glass a reddish color.

Pahoehoe (or ropy lava): Hawaiian term for basaltic lava flows with a ropy surface
- form when the top layer of a lava flow begins to cool. Then, the lower, more fluid lava drags the cooler surface layer into a ropy appearance.
- sometimes the upper layer is glassy, which explains the breaking glass sound you hear as you walk across these lava flows.
- often have large tension cracks form as the lava continues to cool.
- Looking into the walls of these fractures, the surface is often reddish and filled with gas bubble holes.
- These holes were produced by gases escaping from the lava as it was extruded onto the surface.
- The red color is due to the oxidation of the iron bearing minerals by the escaping gases.
- You will probably not see these lavas while using these road logs.

Pele's hair: named after Pele, the Hawaiian goddess of volcanoes. A single strand of Pele's hair has a diameter of less than 0.5 mm (.02 inches) and may be up to 6 feet in length. Pele's hair forms by the stretching or blowing-out of molten basaltic glass from lava fountains, lava cascades, or vigorous lava flows. Pele's hair can be transported high into the air during fountaining, where the wind can carry these glassy strands several tens of miles from a vent.

Phreatic or hydromagmatic eruptions: steam explosions caused when ground water or surface water is heated by magmas or lavas. The intense heat of lavas (as high as 1,170°C for basaltic lava) may cause the water to boil and flash into steam, causing an explosion of steam, water, ash, blocks, and bombs.

Pillow lavas: form when basalt erupts underwater, and mounds of elongate lava "pillows" form by the repeated oozing and quenching of the hot basalt.

- *Continued on the next page.*

Pumice: a gray, glassy, frothy igneous rock that forms during explosive eruptions, or when the top of a lava flow is "whipped" into a glassy foam by escaping gases. This rock will float on water. This rock is also used as an abrasive stone.

Pyroclastic debris: fragments produced by a violent volcanic eruption (*pyro* meaning fiery, and *clastic* meaning broken). Pyroclastic debris includes ash, cinders, block, and bombs.

Pyroclastic flows: an avalanche of hot ash, pumice, rock fragments, and volcanic gas that rushes down the side of a volcano at speeds up to 60 or more miles/hour. The temperature within a pyroclastic flow may be greater than 1000° F.

Rift zones: fractures in the Earth's crust formed when the crust is pulled apart. Hawaii shield volcanoes in generally have three rift zones, along which cinder cones, spatter cones, pit craters, lava flows, lava fountains, ground cracks, and normal faults can form.

Rocks: a naturally occurring assemblage of minerals. However, there are some one mineral rocks like limestone, chert, and dunite.

Scoria: a red, glassy, frothy igneous rock. However, if the rock is gray, glassy and frothy, the rock is called **pumice**. The red color of scoria is due to the oxidation of the iron-bearing minerals present.

Shield volcanoes: big, broad volcanoes with gentle slopes formed by the eruption of fluid basaltic lava. Shield volcanoes commonly have flank eruptions like Kilauea's Mauna Ulu or Puu O oo. The largest volcanoes on Earth are shield volcanoes. This name comes from the volcanoes' resemblance to the shape of a warrior's shield.

Tephra (or pyroclastic debris): refers to fragments of volcanic rock and lava blown into the air by volcanic explosions, or debris carried upward by hot gases in eruption columns or lava fountains. Tephra includes ash, cinders, scoria, pumice, blocks and bombs. The average size of the particles becomes smaller as the distance from the volcano increases. The thickness of the debris also decreases.

Tree molds: form when lava surrounds a tree, solidifies around it, and then the main lava body drains away. The engulfed tree trunks die or are incinerated by the lava leaving cylindrical hollows, or tree molds. Impressed in the basalt, are the impressions of the bark or even the tree rings.

Tsunami: a Japanese word meaning "harbor wave." A tsunami is a wave or a series of waves that form when an ocean basin (or a lake) is shaken by a sudden disturbance that displaces water. Tsunamis are generated by earthquakes, submarine landslides, volcanic eruptions and meteoritic impacts.

Tuff Cone: A type of volcanic cone formed when rising magma comes into contact with seawater or ground water. The resulting steam explosion blasts fine ash size particles into the air, which settles to the ground to form a steep conical hill with a deep wide crater.

Vents: openings in the Earth's crust from which molten rock and volcanic gases are extruded into the air or onto the Earth's surface.

Vesicular basalts:
- contain numerous small gas bubble holes, called vesicles.
- vesicles form when lava is extruded onto the surface, the pressure is reduced and the gases dissolved in the magma escape.
- can be seen as you drive to Haleakala's summit

- *Continued on the next page.*

Volcanoes: conical hills or mountains that form when lavas, gases, and pyroclastic debris is extruded from a central vent.

Active volcanoes: have erupted in historic times.

Dormant volcanoes: volcanoes that have not erupted in history times, but show no erosion of the cone. However, they could erupt again.

Extinct volcanoes: have not erupted in historic time and show erosion of the cone. Thus, extinct volcanoes are considered unlikely to erupt again. Whether a volcano is truly extinct is often difficult to determine. *However,* other sources might give a different definition of "extinct".

Volcanic gases: dissolved gases released into the air when lava is extruded onto the surface. Gases can also be released from subsurface magma through volcanic vents, fumaroles, and hydrothermal systems. The most common gas released by magma is steam (H_2O), then CO_2 (carbon dioxide), SO_2 (sulfur dioxide), (HCl) hydrogen chloride, and other compounds.

THREE COMMON TYPES OF VOLCANOES

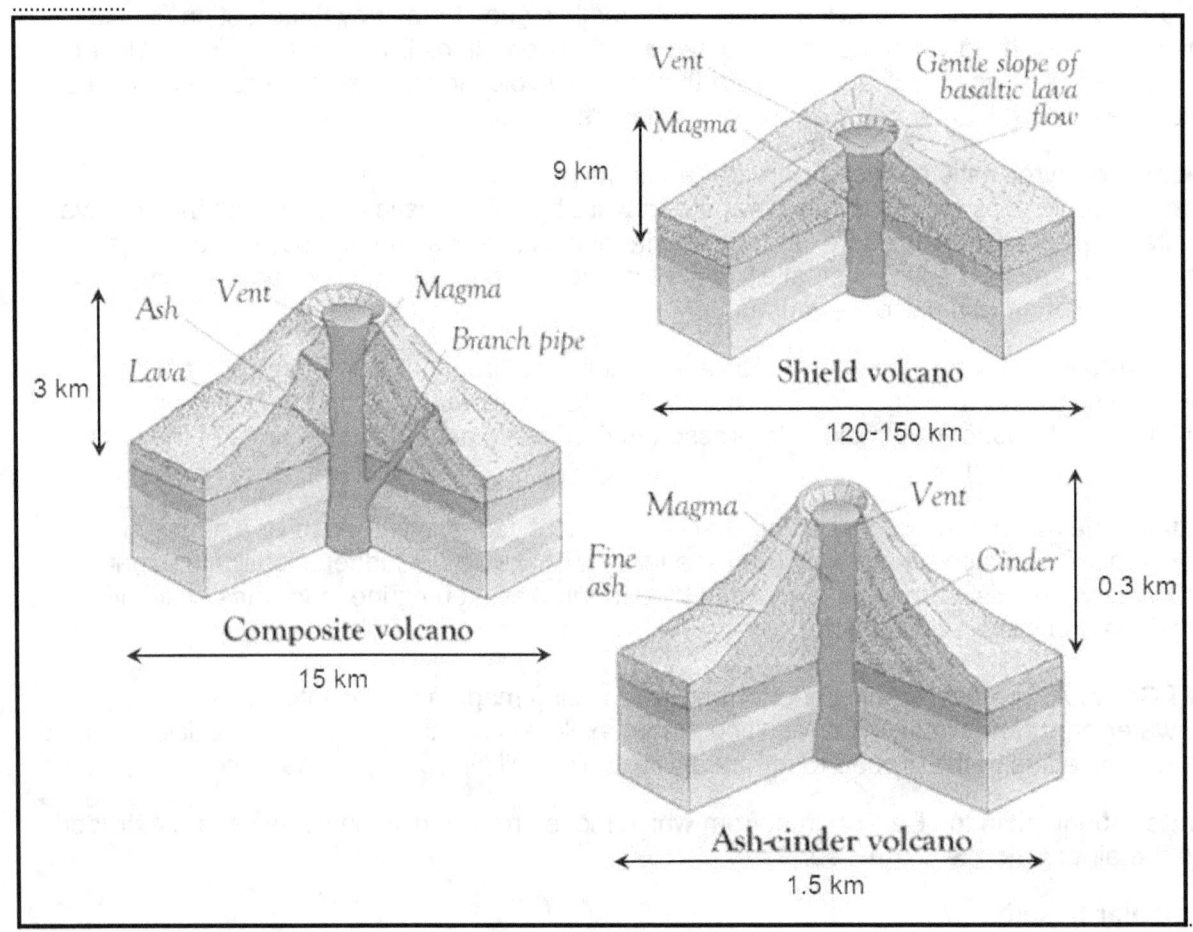

HAWAIIAN PRONUNCIATION GUIDE

The Hawaiian alphabet as only twelve letters:

Five vowels: **a, e, i, o, u**

Seven consonants: **h, k, l, m, n, p, w**

Consonants:

    pronounced the same way as you would say them in English.

    occasionally **w** is sounded as **v.**

Vowels:

- **a:** *"uh"* in unstressed syllables (as in *among*).
  "ah" in stressed syllables (as in *father*).

- **e:** *"eh"* if the e is soft (as in *bet*)
  "ay" when marked as a long e (*as* in *they*).

- **i:** *"ee"* (as in the **i** in *ring or police*).

- **o:** *"oh"* (as in the **o** in *no or vote*).

- **u:** *"oo"* (as in the **u** in *true or too*).

The *glottal stop:*

The glottal stop is an *"unlettered consonant"* which is indicated by a single opening quotation mark ('). This symbol indicated a total break in sound like the pause in "Oh-oh". Example: *pa'u* is PA-oo.

MEANING AND PRONUNCIATION OF MAUI PLACE NAMES

Haleakala: house of the sun.

Hana (ha-na): raining land, low lying sky, or work or profession.

Honomanu (Hoono-ma-noo): shark bay.

Kaanapali: Kaana meaning: *cliff.*

Kahana (ka-ha-na): cutting.

Kahului: (ka-hoo-loo-ee): the winning.

Kalama: torch.

Kamaole: childless.

Kapalua (ka-pa-loo-ah): arms embracing the sea.

Kaupo: night landing (of canoes).

Keanae (Ky-a-nigh): the mullet.

Kihei (key-hay): shawl, cape, tapa garment.

Kipahulu (Key-pah-hoo-loo): worn out soil.

Kula (koo-la): plain, open country, pasture.

Lahaina (la-hi-na): "land of the cruel sun", although the older Hawaiian name is "Lele".

Iao: cloud supreme.

Ma'alaea (ma-ah-lie-ah): red color, as earth.

Mahinahina (ma-hee-na-hee-na): pale moonlight.

Makawao (ma-ka-wow): forest beginning.

Makena: abundance.

Maui (mow-wee): the Demigod that snared the sun.

Molokini: many ties.

Napili (na-pee-lee): the joinings or *pili* grass.

Olowalu (oh-low-wah-loo): many hills.

Paia (pa-ee-ah): noisy.

Pukalani (poo-ka-law-nee): hole in the heaven (clouds).

Pua a Kaa State Park (Poo-ah-ah Kah-ah): Park of the rolling pigs.

Puunene: goose hill.

Waianapanapa State Park (Why-ah-nah-pa-nah-pa): glistening water.

Wailea (why-lay-ah): water (of) Lea, goddess of canoe makers.

Wailuku (why-loo-koo): water of destruction.

Hawaiian Words You Should Know

Alii: chief, royalty.

Aloha: Aloha mean hello, goodbye, and greetings as a warm welcome.

Hale: house.

Heiau: temple.

He mea iki: you are welcome.

Haole, Ma Li Hini, and Kama'aina: someone who is not of Hawaiian descent, may be used in an unflattering way.

Hono: bay

Imu: earthen oven used at a luau.

Kahuna: priest

Kai: seaward

Kamaaina: hawaiian native

Kane: man, used in restrooms

Kapu: taboo

Keiki: children

Koa: type of hard wood

Kona: leeward or a leeward wind

Lae: geographic point or headland

Lei: garland of flowers

Limu: seaweed

Luau: traditional feast

Mahalo: thank you.

Makai: the ocean side or direction towards the sea. You'll also hear windward for the eastern and wet part of an island and leeward for the western and dry part of an island.

Mahimahi: dophin fish (not a dophin)

Ma Li Hini: a tourist.

Mana: supernatural power

Mano shark

Maui no ka oi: Maui is the best

Mauka: means inland, towards the mountains, to go toward the inner parts of an island, away from the ocean.

Mauna: mountain

Mele: song

Mele Kalikimaka: Merry Christmas

Moana: open sea, ocean

Muumuu: long, loose-fitting dress introduced by missionaries

Nene: hawaiain goose

Nui: large

Ohana: family

Ono: good, delicious

Pali: cliff

Paniolo: Hawaiian cowboy

Poi: starchy paste made for taro roots

Poke: cubed, marinated, spiced raw fish

Pono: honest

Pua: flower

Pupu: appetizers found at luaus.

Puu: hill

Shaka: An expression as much as a term,

Shaka exemplifies island spirit. Middle fingers remain in a fist and thumb and pinky are extended and wiggled, shaka is hang loose, a friendly gesture.

Wahine: woman

Wai: fresh water

Wailele: water fall

Wikiwiki: the Honolulu airport shuttle, is the ground transportation between terminals, means speedy.

Kama'aina: a local island resident, native or someone who now lives on the island

MISPRONOUNCED HAWAIIAN WORDS

Likelike
Likelike is often referred to as "Like Like" by visitors, but it is pronounced: "**Lee-kay-Lee-kay**."

Humuhumunukunukuapuaa
This word can be pronounced correctly by most Hawaii residents and those who study the syllables: "**Hoo-moo-hoo-moo-noo-koo-noo-koo-ah-poo-ah-ah**." The Hawaiian name translates as the "fish with a nose like a pig."

Hanauma Bay
The correct pronunciation is "**Ha-now-mah**."

Lanai
If you're talking about the island of Lanai, it's pronounced "**La-nah-ee**." If you're talking about a porch or veranda, the word can flow out ("lah-nai") without stops or pauses.

HAWAIIAN PRONUNCIATIONS

TO HEAR MAUI PLACE NAMES PRONOUNCED, GO TO:

http://hawaiian-words.com/hawaii-place-names

TO HEAR HAWAIIAN WORDS PRONOUNCED, GO TO:

http://hawaiian-words.com/common

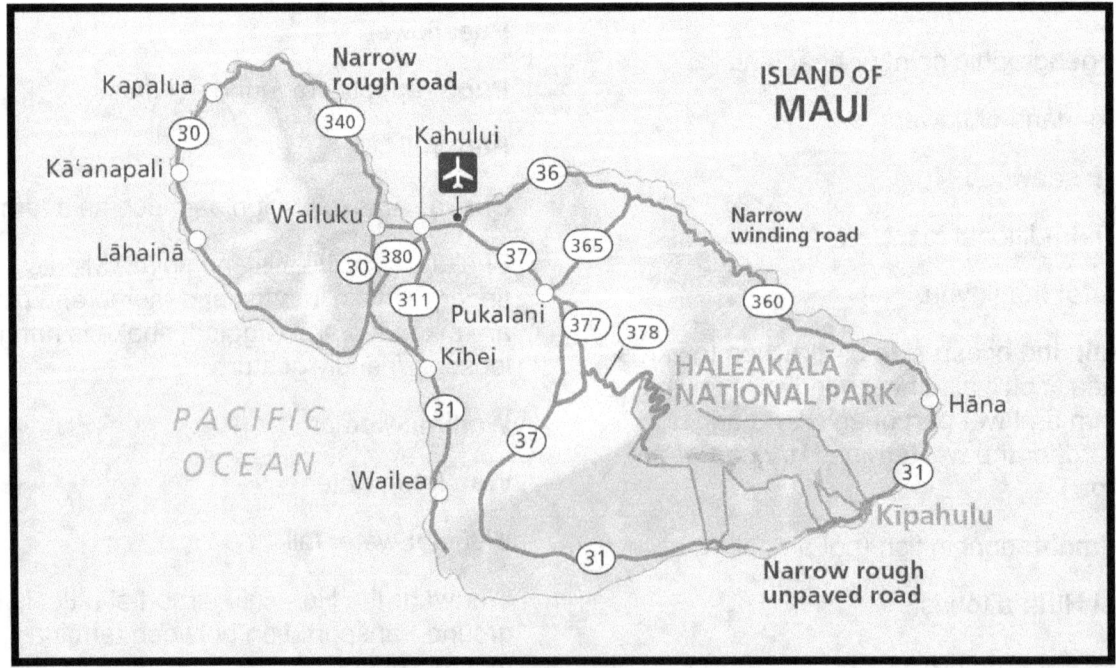

Courtesy of the United States Geological Survey

WEBSITES FOR NEWS AND INFORMATION

The Maui News: Maui's daily newspaper: **mauinews.com**

The Lahaina News: Weekly newspaper with news for West Maui: **lahainanews.com**

Maui Weather Today: Three times a week, a Maui weatherman gives weather forecasts and comments on life on Maui to the Internet. This site has several links to satellite shots of the islands: **hawaiiweathertoday.com**

Maui Now: Up-to-the minute coverage of what's happening on Maui. Plus: events, reviews, weather, sports, business, and more: **mauinow.com**

Maui Time: Maui News covering best of Maui, things to do in Maui, restaurant reviews, places to eat, entertainment, politics and more: **mauitime.com**

News History of Hawaii: Summarizes the history of the island of Maui: **gohawaii.com/maui/about/history**

Maui Museum: mauimuseum.org

Best of Maui Guide: Lahaina Through the Ages: **bestofmauiguide.com/lahainahistory**

LIVE CAMERAS

Maui Webcams: Webcam shows scenic viewpoints of Maui: **mauihawaii.org/webcams.htm**

Maui Live Surf Cam - Kaanapali - Lahaina - Wailea - Kapalua:

livesurfcamhawaii.com/maui/maui.htm

NINE IMPORTANT TIPS FOR HIKING IN HAWAII

1. **Don't keep valuables in your car:** Many hikes require you to park your car in remote locations. Theft in those situations is not uncommon.

2. **Don't hike alone.** It is easy to make a wrong turn and become disoriented when hiking by yourself. Only hike on trails you know you can handle.

3. **Be back by nightfall.** Since the Islands are close to the equator, it gets dark very quickly once the sun sets.

4. **Watch where you swim.** Many Hawaii streams and waterfall pools contain an infectious bacteria called *Leptospirosis*, which can be dangerous.

5. **Water.** It's warm in Hawaii. Don't wait to start drinking until you are thirsty. Drink fluids before and throughout your hike in order to prevent dehydration.

6. **Food.** Bring snacks like fruit, a protein bar, or trail mix to keep up your energy.

7. **Hiking shoes.** Once you get off cleared trails, you may need to climb narrow ridges or rocks. This may come as a disappointment, but your Crocs will not make the cut.

8. **A windbreaker.** Rainstorms and high winds can quickly *spring-up.* Don't be on a mountain peak, soaked and freezing without a sweatshirt or light windbreaker.

9. **Sunscreen.** Hawaiian sun can be extremely harsh, especially on skin that hasn't seen UV rays since last summer. Use sun screen at all times.

TYPES OF HAWAII LAVAS

Ropy or pahoehoe basalts:
- form when the top layer of a lava flow begins to cool. Then, the lower, more fluid lava drags the cooler surface layer into a ropy appearance.
- sometimes the upper layer is glassy, which explains the breaking glass sound you hear as you walk across these lava flows.
- often have large tension cracks form as the lava continues to cool.
- Looking into the walls of these fractures, the surface is often reddish and filled with gas bubble holes.
- These holes were produced by gases escaping from the lava as it was extruded onto the surface.
- The red color is due to the oxidation of the iron bearing minerals by the escaping gases.
- You will probably not see these lavas while using these road logs.

Blocky or aa basalts:
- form from a slower, cooler, and hence more viscous lava flow advances across a landscape.
- the forward movement of this slow moving lava flow causes the cooled upper crust to break into irregular shaped blocks that tumble down the front of this slow moving lava flow.
- are difficult to walk over, so please use caution walking across a blocky lava.
- often contain gas bubbles, called vesicles, produced by gases escaping from the lava when it was extruded onto the surface.
- can be seen on the south side of Haleakala.

Vesicular basalts:
- contain numerous small gas bubble holes, called vesicles.
- vesicles form when lava is extruded onto the surface, the pressure is reduced and the gases dissolved in the magma escape.
- can be seen as you drive to Haleakala's summit

Pele's hair:
- form when lava fountains and steam explosions stretches lava into long strands of glassy filaments.
- is a natural spun glass that has the look, feel, and texture of blonde human hair.
- not see while using these road logs.

Limu O Pele
- forms when lava flows into the sea and generates steam to produce a huge bubble of lava.
- develops as this bubble grows larger, and the lava bubble walls thin and solidify.
- when the bubble bursts, paper-thin glassy flakes are produced that look like pieces of seaweed (Limu).
- bubble flakes are called *Limu O Pele*.
- Also not seen while using these road logs.

APPENDIX ONE

MAUI GEOLOGIC MAPS

The following geologic maps were taken from selected portions of the:

Geologic Map of the State of Hawaii, Sheet 7 -Island of Maui
by
Sherrod, D.R, Sinton, J.M, Watkins, S.E. and Brunt, K.M., 2007

NOTE: The following maps have been saved as jpeg files. To make some of the maps more readable, the jpeg file have been enlarged or stretched. Thus, parts of the map have been distorted for easier readability of the formation names.

STRATIGRAPHIC AGES OF THE VOLCANIC UNITS ON MAUI

Hana Volcanics: Qhn6: 0-1500 years ago to Qhn0: 50,000 to 140,000 years ago

Kula Volcanics: 0.93 to 0.15 million years ago

Honomanu Basalt: 1.1 to 0.97 million years ago.

Lahaina Volcanics: 0.6 to 0.3 million years ago

Honolua Volcanics atop Wailuku: 1.3 to 1.1 million years ago.

Wailuku Basalt: 2.0 to 1.3 million years ago.

From: **Haleakalā National Park,** *Geologic Resources Inventory Report* Natural Report NPS/NRSS/GRD/NRR—2011/453

SYMBOLS USED ON MAUI GEOLOGIC MAPS FOR *SEDIMENTS*

Qa - Alluvium (Holocene)

Qbd - Beach deposits (Holocene)

Ody Younger dune deposits (Holocene)

Odo Older beach deposits (Holocene and Pleistocene)

QTao - Older alluvium (Pleistocene and Pliocene)

SYMBOLS USED ON MAUI GEOLOGIC MAPS FOR *HALEAKALA VOLCANICS*

Qhn- Hana Volcanics (Holocene and Pleistocene)

Qkul - Kula Volcanics; lava flows (Pleisticene)

SYMBOLS USED ON MAUI GEOLOGIC MAPS FOR THE *WEST MAUI VOLCANO*

Qlhl Lahaina Volcanics (Pleistocene)

Qul Honolua Volcanics (Pleistocene)

Qtwl Wailuku Basalt (Pleistocene and Pliocene?)

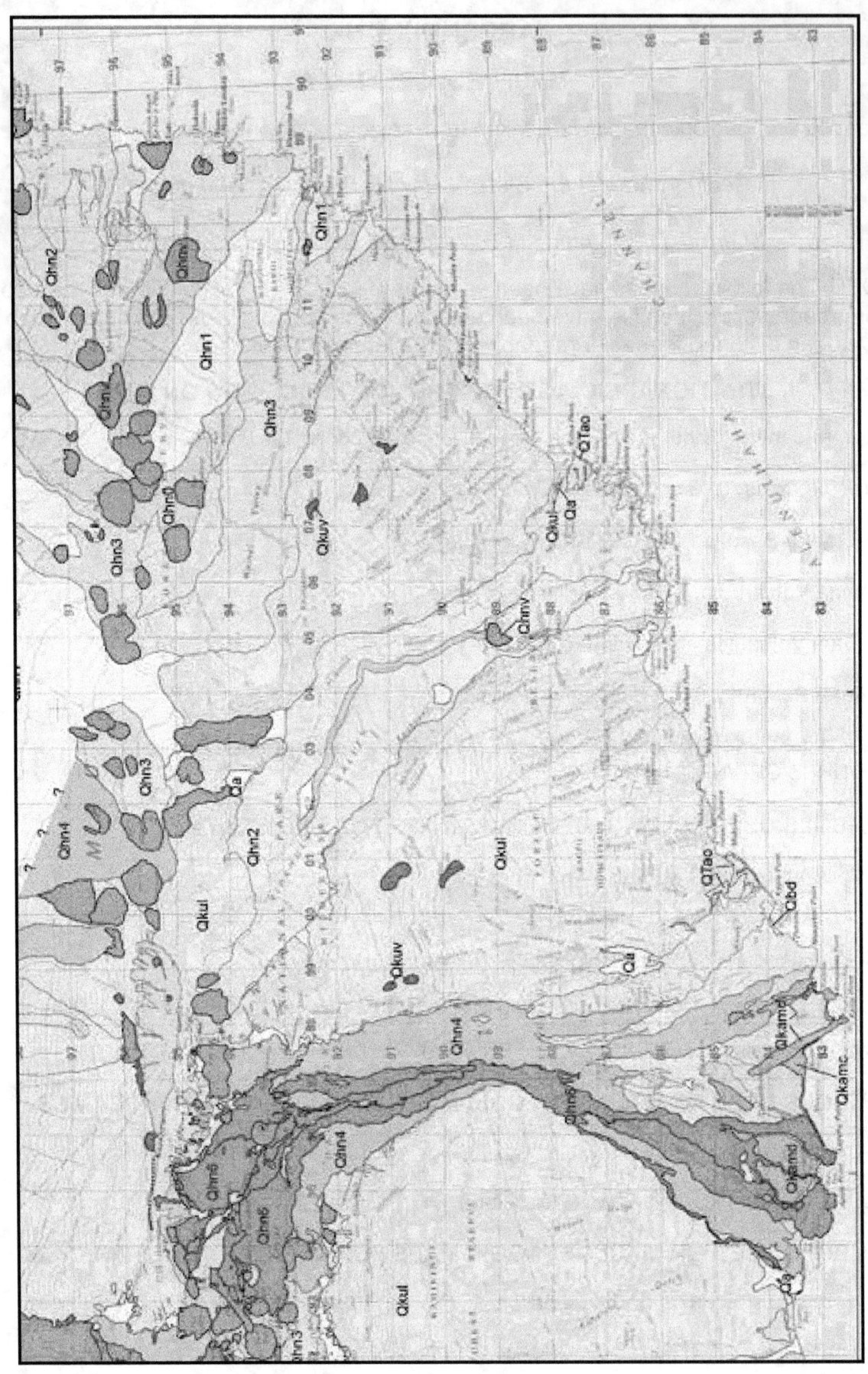

GEOLOGIC MAP OF SOUTH EASTERLY MAUI

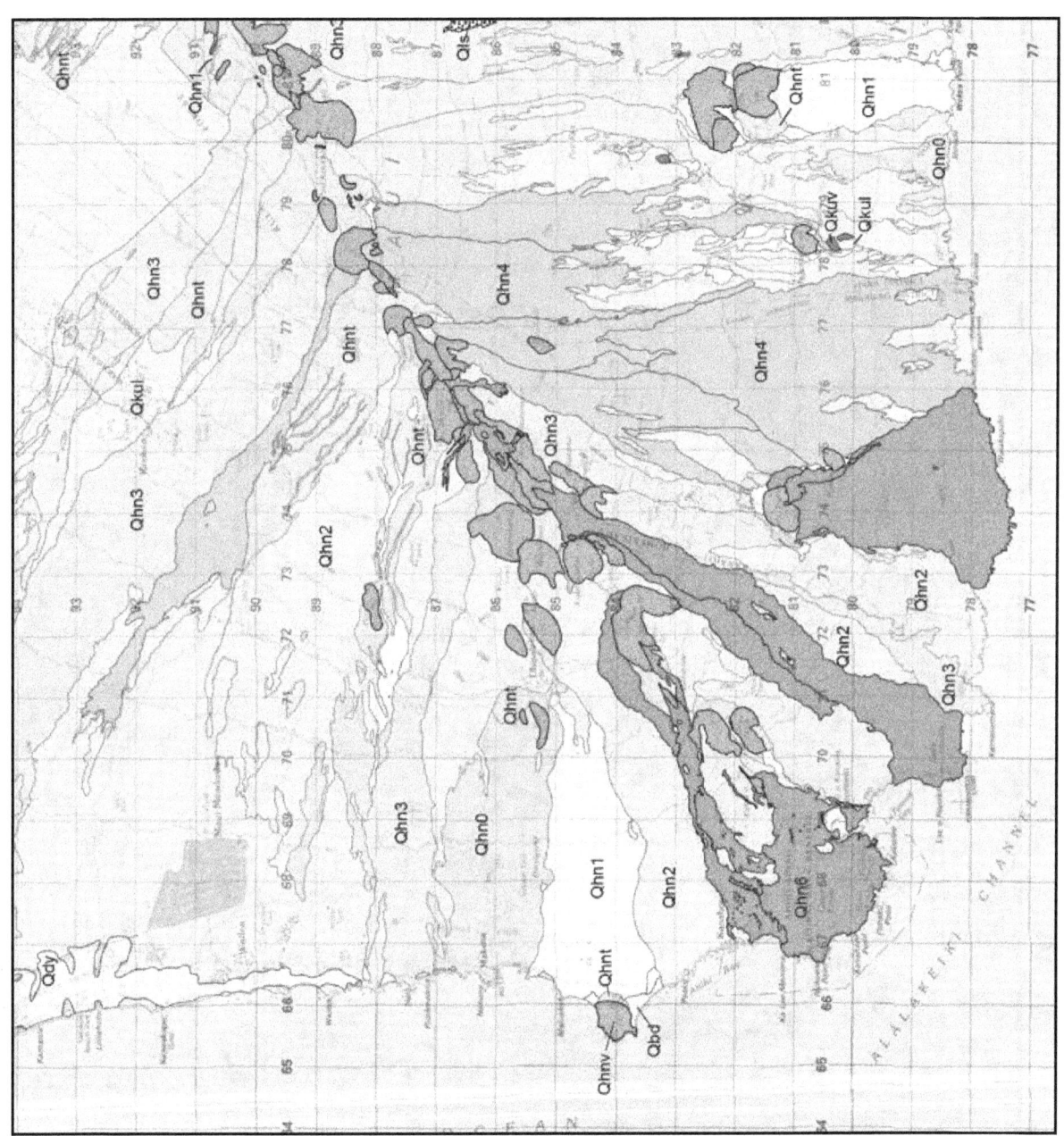

GEOLOGIC MAP OF SOUTH CENTRAL MAUI

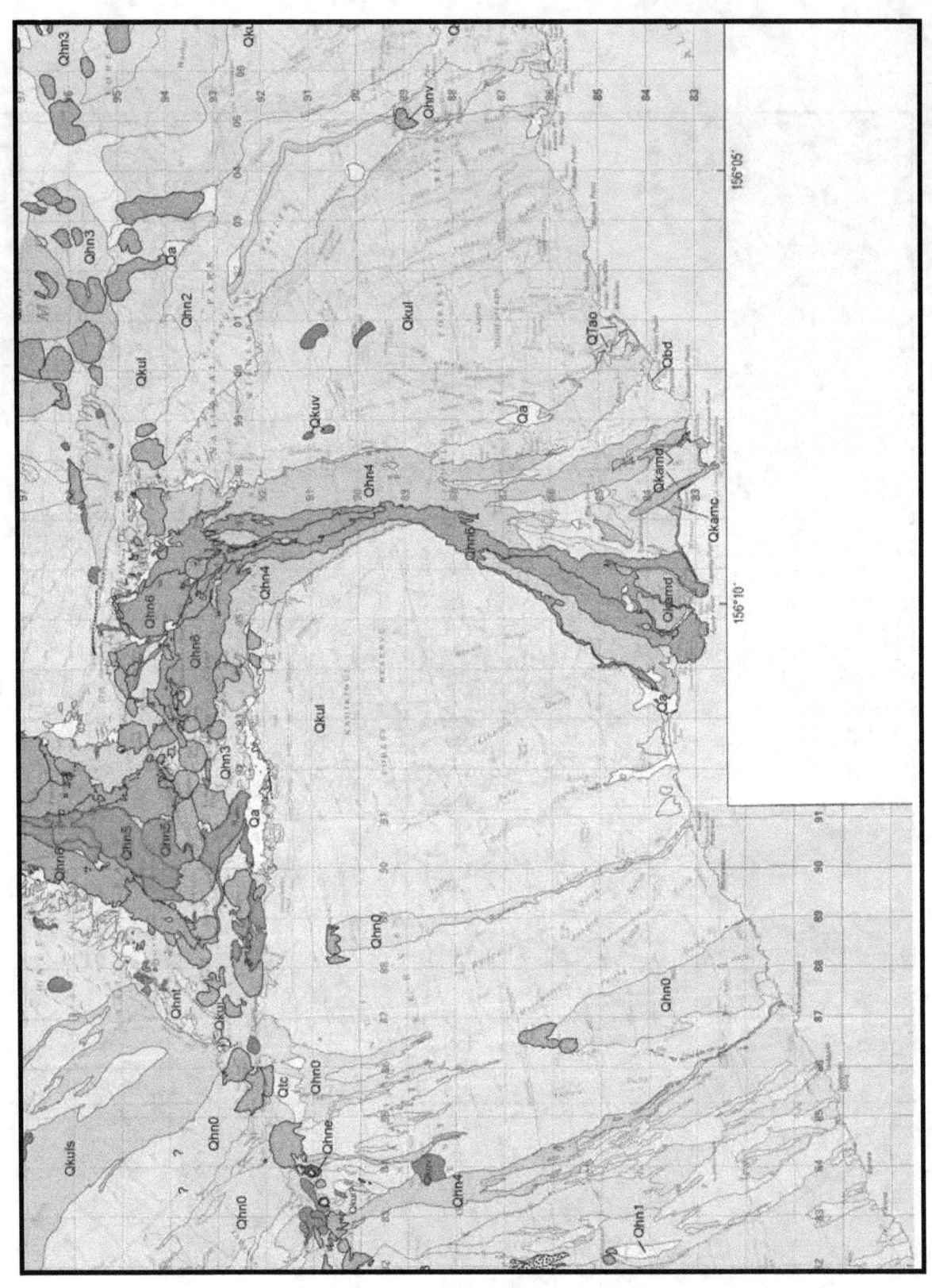

GEOLOGIC MAP OF CENTRAL SOUTHEASTERLY MAUI

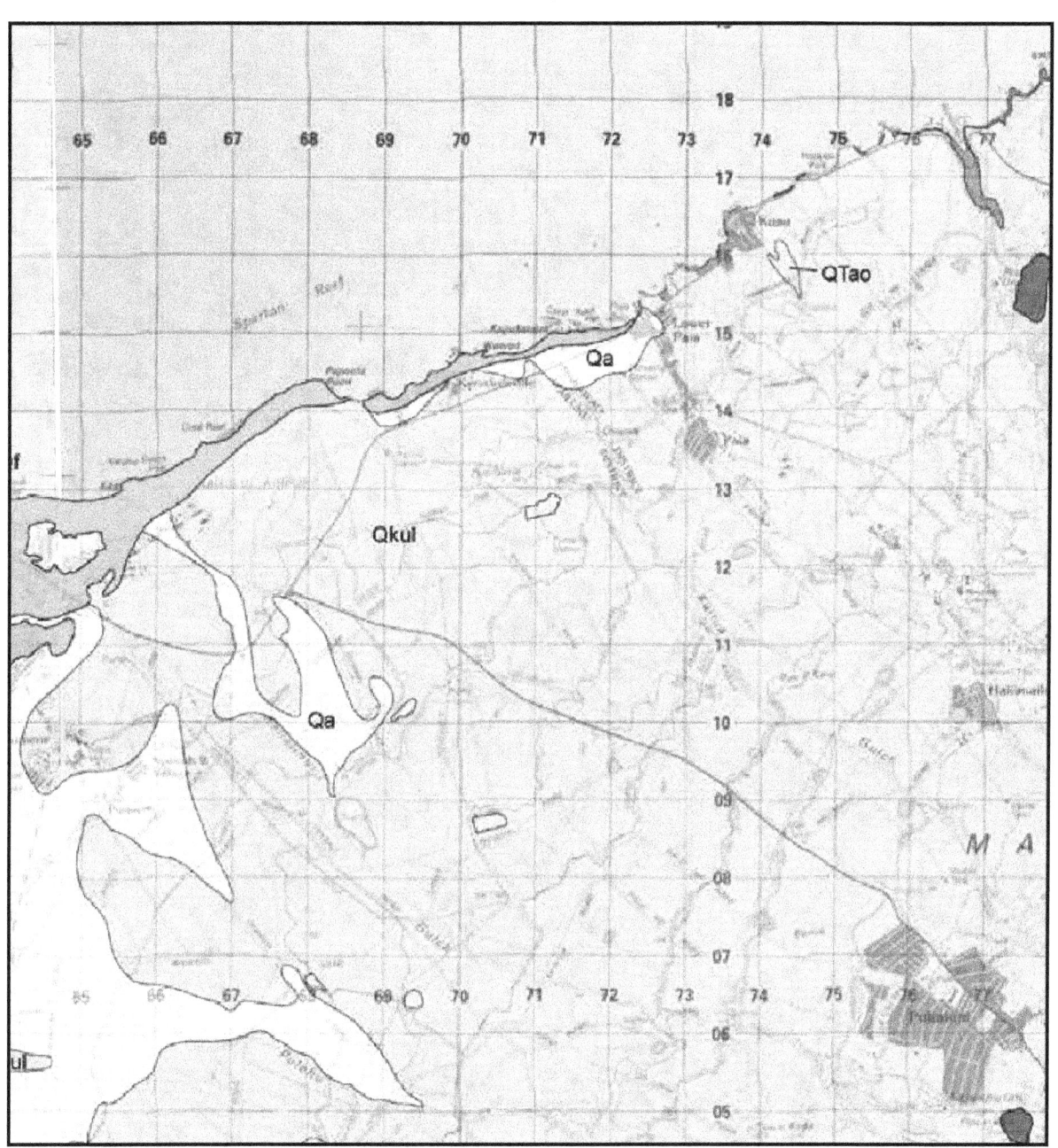

GEOLOGIC MAP OF NORTH EAST MAUI

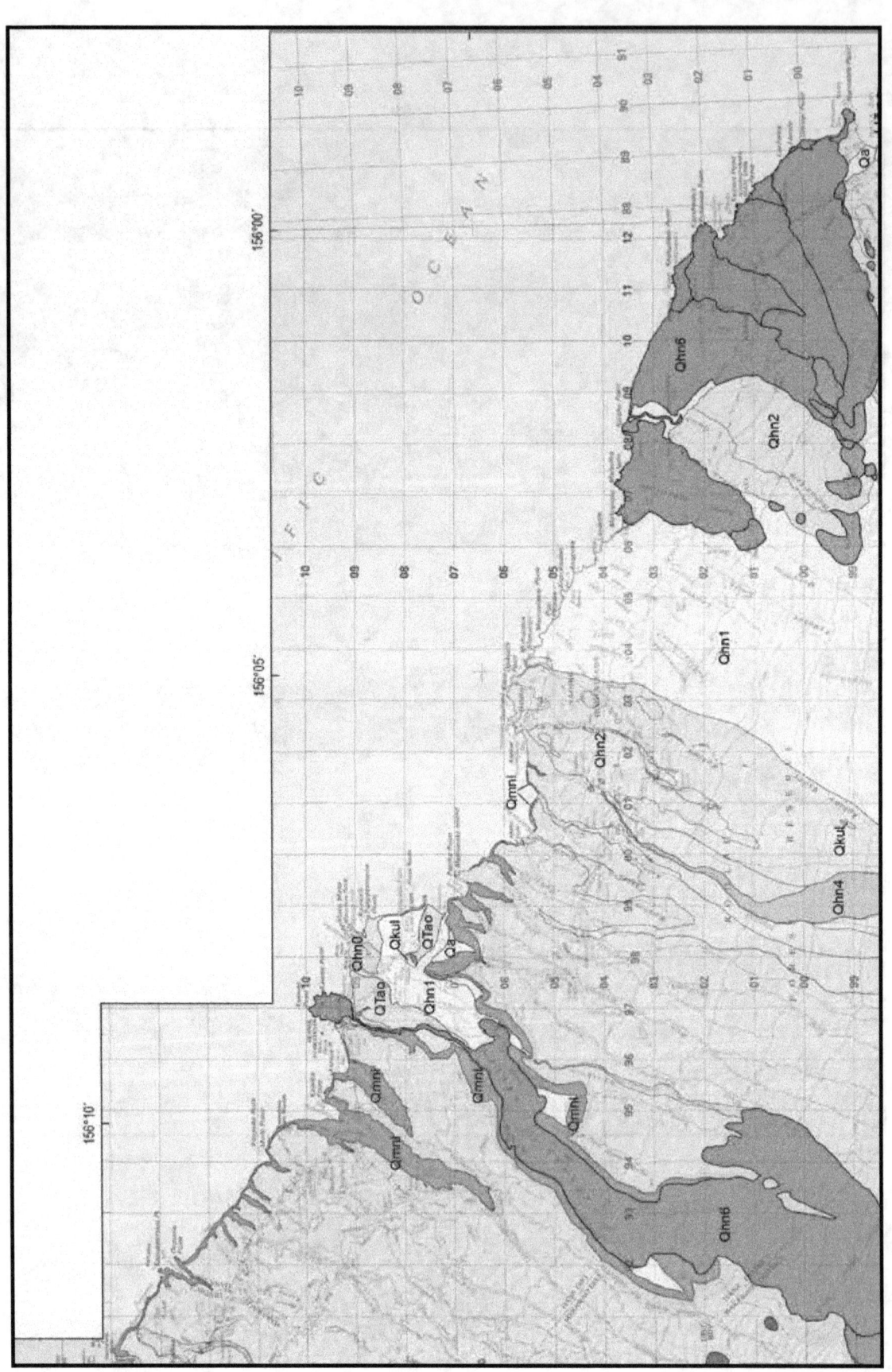

GEOLOGIC MAP OF NORTH EASTERLY MAUI

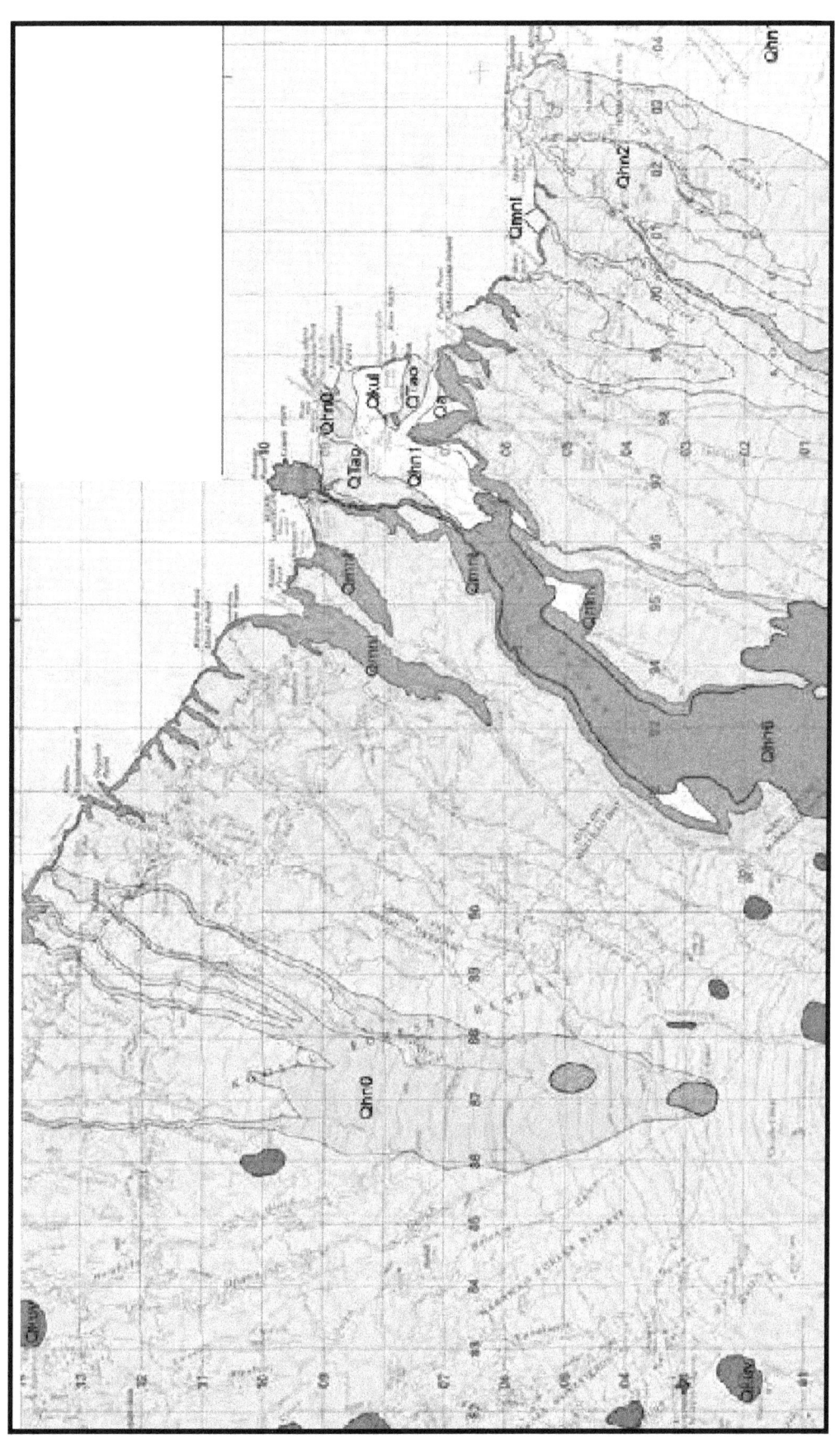

GEOLOGIC MAP OF EAST MAUI

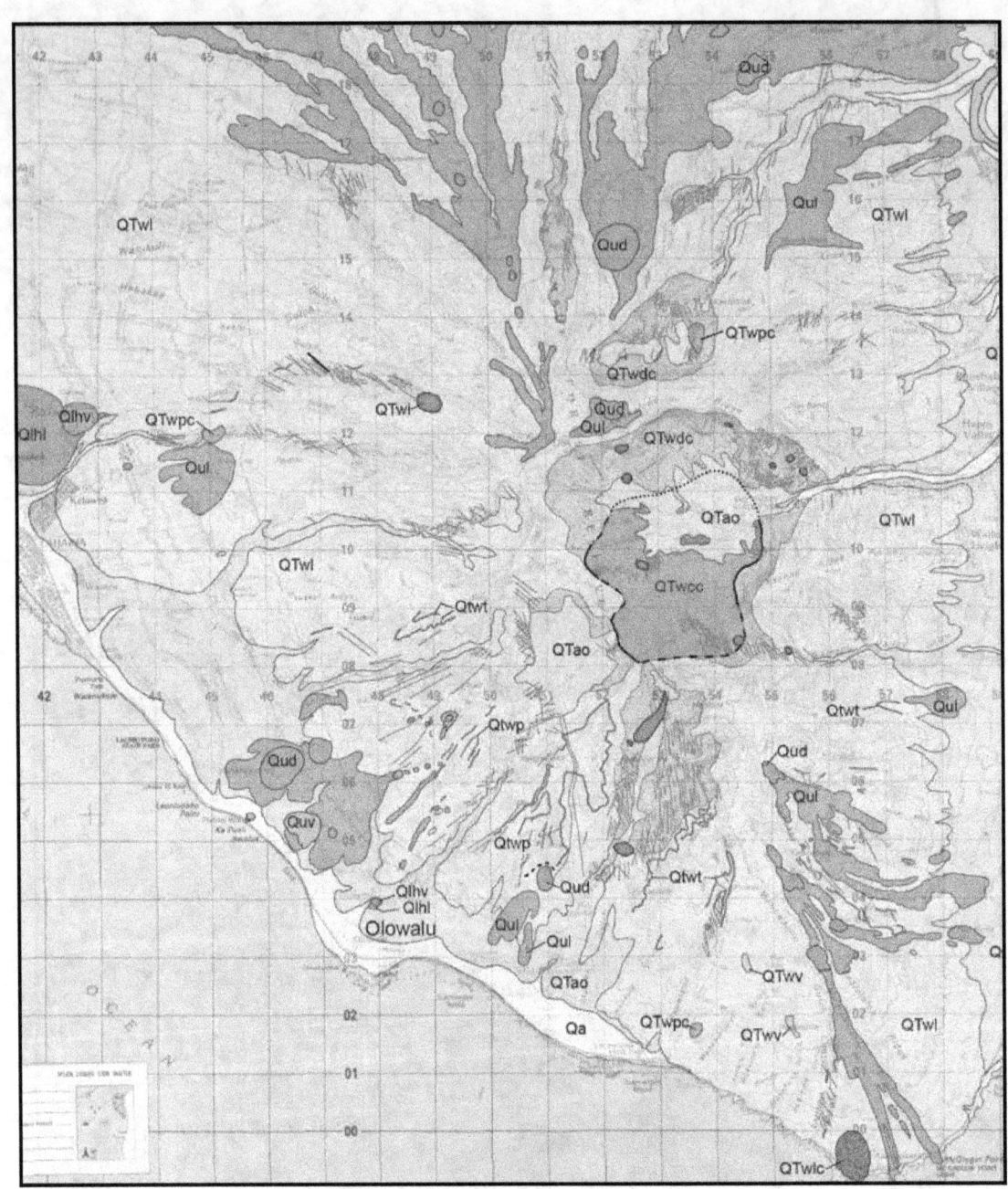

GEOLOGIC MAP OF NORTHWESTERN MAUI

REFERENCES USED IN THIS BOOK

Clark, John, The Beaches of Maui County, 1989.
Decker, Robert, and Decker, Barbara, Road Guide to Haleakala and the Hana Highway, 1992.
Doughty, A., Maui Revealed, 2012
Hazlett, R. W., Hyndman, D. W., Road Side Geology of Hawaii, 1996.
Kirch P. V., Hartshor A. S., Chadwic O. A., Vitouse P. M., Sherrod D. R., Coil J., Holm L., and.. Sharp, D. W., 2004, Environment, agriculture, and settlement patterns in a marginal Polynesian Landscape.
Kyselka, W. and Lanterman, R, Maui How It Came To Be, 1980.
Porter, S. C., M. Stuiver, and I. C. Yang. 1977. Chronology of Hawaiian Glaciations. Science 195(4273):61–62.
Sinton, John, Maui Field Guide, 2006.
Thornberry-Ehrlich, T. 2011. Haleakalā National Park: geologic resources inventory report. Natural Resource Report NPS/ NRSS/GRD/NRR—2011/453. National Park Service, Ft. Collins, Colorado.
Haleakalā National Park Geologic Resources Inventory Report Natural Resource Report NPS/NRSS/GRD/NRR—2011/453
A Natural History of the Hawaiian Islands: Selected Readings II edited by E. Alison Kay, 1994
mauiguidebook.com › Central Maui › Adventures & Sights
http://www.nature.org/ourinitiatives/regions/northamerica/unitedstates/indiana/journeywithnature/kudzu-invasive-   species.xml
http://hawaiian-words.com/hawaii-place-names
http://gohawaii.about.com/od/mauiphotos/ig/Maui-Beach-Photos/honokowai_002.htm
http://www.world-of-waterfalls.com/hawaii-maui.html
http://www.hawaii-guide.com/maui/articles/maui_waterfalls
http://www.hawaii-guide.com/maui/articles/maui_waterfalls
https://en.wikipedia.org/wiki/Hosmer%27s_Grove
http://online.wsj.com/news/articles/SB10001424052748704320104575015431045157378
http://gohawaii.about.com/od/maui/p/maalaea_profile.htm
http://www.gohawaii.com/maui/regions-neighborhoods/upcountry-maui/kula
http://www.to-hawaii.com/maui/cities/kula.php
http://www.hawaiiweb.com/maalaea-bay-and-harbor-maui.html
http://en.wikipedia.org/wiki/Lahaina,_Hawaii
http://www.hort.purdue.edu/newcrop/crops/Sugar_cane.html
http://www.fws.gov/refuge/kealia_pond/
http://kiheibeachresorts.com/beaches.html
http://www.actionsportsmaui.com/beaches.html
http://mauiguidebook.com/beaches/kalepolepo-beach-park/
http://www.fodors.com/world/north-america/usa/hawaii/maui/review-414853.html
http://www.hawaiisharks.com/incidents.html
http://en.wikipedia.org/wiki/La_Perouse_Bay
http://www.frommers.com/destinations/maui/662916
http://en.wikipedia.org/wiki/La_Perouse_Bay
http://hawaiitrails.ehawaii.gov/trail.php?TrailID=MA+18+001
http://www.roadtohana.info/road-to-hana-history-kings-road/
http://www.mauimagazine.net/Maui-Magazine/Fall-2002/Along-the-Kings-Trail/
http://www.soest.hawaii.edu/coasts/publications/GeologyofHawaiiReefs.pdf   has rain fall map
http://pubs.usgs.gov/sir/2012/5010/sir2012-5010.pdf
http://hvo.wr.usgs.gov/volcanowatch/archive/2002/02_06_27.html
http://myhawaii.biz/maui-geological-history/
http://www.gohawaii.com/maui/about/history
http://www.bestofmauiguide.com/lahainahistory.html
http://bulletin.geoscienceworld.org/content/115/6/683

http://www.hawaiiweb.com/maui-tropical-plantation-and-country-store-maui.html
http://www.tropicalgardensofmaui.com/index.aspx
http://home.howstuffworks.com/lawn-garden/professional-landscaping/basics/monkey-pod-tree2.htm
http://alohaisles.com/maui/sugar_cane.html
http://www.coastalliving.com/travel/top-10-spots-to-snorkel-00400000047380/page12.html
http://en.wikipedia.org/wiki/Lahaina,_Hawaii
http://www.hort.purdue.edu/newcrop/crops/Sugar_cane.html
http://www.fws.gov/refuge/kealia_pond/
http://kiheibeachresorts.com/beaches.html
http://www.actionsportsmaui.com/beaches.html
http://mauiguidebook.com/beaches/kalepolepo-beach-park/
http://www.fodors.com/world/north-america/usa/hawaii/maui/review-414853.html
http://www.hawaiisharks.com/incidents.html
http://en.wikipedia.org/wiki/La_Perouse_Bay
http://www.frommers.com/destinations/maui/662916
http://en.wikipedia.org/wiki/La_Perouse_Bay
http://hawaiitrails.ehawaii.gov/trail.php?TrailID=MA+18+001
http://www.roadtohana.info/road-to-hana-history-kings-road/
http://www.mauimagazine.net/Maui-Magazine/Fall-2002/Along-the-Kings-Trail/
http://www.soest.hawaii.edu/coasts/publications/GeologyofHawaiiReefs.pdf
the above source has rain fall map
http://pubs.usgs.gov/sir/2012/5010/sir2012-5010.pdf
http://hvo.wr.usgs.gov/volcanowatch/archive/2002/02_06_27.html
http://myhawaii.biz/maui-geological-history/
http://www.gohawaii.com/maui/about/history
http://www.bestofmauiguide.com/lahainahistory.html
http://bulletin.geoscienceworld.org/content/115/6/683
http://www.hawaii-guide.com/maui/articles/maui_weather
http://www.soest.hawaii.edu/coasts/publications/shores/6Streams_and_rockfalls_FLETCHER-final.pdf
http://hvo.wr.usgs.gov/volcanowatch/archive/2001/01_09_13.html
http://www2.hawaii.edu/~nasir/0811MauiGeology.ppt#257,1,Slide 1
http://volcano.oregonstate.edu/vwdocs/volc_images/north_america/hawaii/Kahoolawe_map.gif
http://www.gohawaii.com/molokai/about/history
http://www.gohawaii.com/maui/regions-neighborhoods/west-maui/kapalua
http://www.hawaiigaga.com/maui/beaches/dt-fleming-beach-park.aspx
http://www.gohawaii.com/maui/regions-neighborhoods/west-maui/honolua-bay
http://tps.cr.nps.gov/nhl/detail.cfm?ResourceId=192&ResourceType=
http://www.asce.org/People-and-Projects/Projects/Landmarks/East-Maui-Irrigation-System/
http://www.frommers.com/destinations/maui/662935
http://www.discoverhawaiitours.com/attractions/wailua-falls/
http://laainakai.com/nahiku_roadside
http://energy.hawaii.gov/wp-content/uploads/2011/10/Update-of-the-Statewide-Geothermal-Resource-Assessment-of-Hawaii.pdf
http://temptationtours.com/science-city-high-atop-mauis-10000-foot-mt-haleakala/
http://www.haleakala.national-park.com/info.htm
http://www.pnas.org/content/101/26/9936.full.pdf
http://www.hawaiilife.com/articles/2012/01/hawaiis-winds/
http://www.world-guides.com/north-america/usa/hawaii/maui/maui_weather.html
http://www.sciencedirect.com/science/article/pii/S0377027302003852
http://www.kulacatholiccommunity.org/history.html
http://www.to-hawaii.com/maui/waterfalls/hanawifalls.php
http://www.unrealhawaii.com/wp-content/uploads/2013/08/Map-and-Descriptions.pdf

http://www.mauihawaii.org/maui-questions/lahaina-parking.htm
http://www.visitorinfohawaii.com/the-history-of-kaanapali
http://en.wikipedia.org/wiki/Olowalu,_Hawaii
http://hvo.wr.usgs.gov/volcanowatch/archive/2003/03_04_10.html
http://www.hawaiiguideme.com/2011/09/12/hono-a-piilani-%E2%80%93-the-bays-of-piilani/
http://www.hawaii-guide.com/maui/sights/oheo_gulch_kipahulu
http://www.to-hawaii.com/maui/cities/keokea.php
http://online.wsj.com/news/articles/SB10001424052748704320104575015431045157378
http://hvo.wr.usgs.gov/volcanowatch/archive/2001/01_02_08.html
http://www.fourwindsmaui.com/molokini-crater
https://plus.google.com/+OlowaluGeneralStoreLahaina/about?gl=us&hl=en
http://www.lighthousefriends.com/light.asp?ID=917
http://tourmaui.com/maui-ancient-history/
http://pacificscience.files.wordpress.com/2014/02/pac-sci-early-view-68-3-4.pdf
http://www.roadtohana.info/
https://www.tombarefoot.com/hawaii-information/macgregor-point-lighthouse/418
http://www.pacifichorticulture.org/articles/african-tulip-tree-2/
http://hcsugar.com/what-we-do/collecting-and-delivering-water/
http://hbs.bishopmuseum.org/good-bad/uluhe.html
http://www.cr.nps.gov/history/online_books/kona/historyt.htm
http://www.asce.org/People-and-Projects/Projects/Landmarks/East-Maui-Irrigation-System/
http://gohawaii.about.com/od/maui/p/maalaea_profile.htm
http://www.gohawaii.com/maui/regions-neighborhoods/upcountry-maui/kula
http://www.to-hawaii.com/maui/cities/kula.php
http://www.hawaiiweb.com/maalaea-bay-and-harbor-maui.html
http://www.hawaiiweb.com/maui-tropical-plantation-and-country-store-maui.html
http://www.tropicalgardensofmaui.com/index.aspx
http://home.howstuffworks.com/lawn-garden/professional-landscaping/basics/monkey-pod-tree2.htm
http://alohaisles.com/maui/sugar_cane.html
http://www.coastalliving.com/travel/top-10-spots-to-snorkel-00400000047380/page12.html
http://en.wikipedia.org/wiki/Lahaina,_Hawaii
http://www.hort.purdue.edu/newcrop/crops/Sugar_cane.html
http://www.fws.gov/refuge/kealia_pond/
http://kiheibeachresorts.com/beaches.html
http://www.actionsportsmaui.com/beaches.html
http://mauiguidebook.com/beaches/kalepolepo-beach-park/
http://www.fodors.com/world/north-america/usa/hawaii/maui/review-414853.html
http://www.hawaiisharks.com/incidents.html
http://en.wikipedia.org/wiki/La_Perouse_Bay
http://www.frommers.com/destinations/maui/662916
http://en.wikipedia.org/wiki/La_Perouse_Bay
http://hawaiitrails.ehawaii.gov/trail.php?TrailID=MA+18+001
http://www.roadtohana.info/road-to-hana-history-kings-road/
http://www.mauimagazine.net/Maui-Magazine/Fall-2002/Along-the-Kings-Trail/
http://www.soest.hawaii.edu/coasts/publications/GeologyofHawaiiReefs.pdf. The above source has rain fall map
http://pubs.usgs.gov/sir/2012/5010/sir2012-5010.pdf
http://hvo.wr.usgs.gov/volcanowatch/archive/2002/02_06_27.html
http://myhawaii.biz/maui-geological-history/
http://www.gohawaii.com/maui/about/history
http://www.bestofmauiguide.com/lahainahistory.html
http://bulletin.geoscienceworld.org/content/115/6/683
http://www.hawaii-guide.com/maui/articles/maui_weather

http://www.soest.hawaii.edu/coasts/publications/shores/6Streams_and_rockfalls_FLETCHER-final.pdf
http://hvo.wr.usgs.gov/volcanowatch/archive/2001/01_09_13.html
http://www2.hawaii.edu/~nasir/0811MauiGeology.ppt#257,1,Slide 1
http://volcano.oregonstate.edu/vwdocs/volc_images/north_america/hawaii/Kahoolawe_map.gif
https://mauiguide.com/wp-content/uploads/edd/2017/09/maui-guide-ebook-1.pdf
http://www.kauai.com/fern-grotto
www.hawaiistateparks.org/parks/maui/index.cfm?park_id=36
www.gohawaii.com/.../iao-valley-state-p...
www.gohawaii.about.com/od/maui/ss/iao_valley_2.htm
www.hawaiiweb.com
mauiguidebook.com › Central Maui › Adventures & Sights
www.co.maui.hi.us/facilities/Facility/Details/103
www.hawaiiweb.com/kepaniwai-heritage-gardens.html
www.fodors.com › Destinations › USA › Hawaii › Maui › Sights
www.hawaiistateparks.org/parks/maui/index.cfm?park_id=36
www.gohawaii.com/.../iao-valley-state-p...
www.gohawaii.about.com/od/maui/ss/iao_valley_2.htm
www.hawaiiweb.com
www.co.maui.hi.us/facilities/Facility/Details/103
www.hawaiiweb.com/kepaniwai-heritage-gardens.html
www.fodors.com › Destinations › USA › Hawaii › Maui › Sights
www.lonelyplanet.com/.../park/.../parks-gardens/kepaniw
www.gogobot.com › ... › Hawaii › Maui › Wailuku › Things to do
www.gohawaii.com/maui/...maui/iao-val...
www.lonelyplanet.com/.../park/.../parks-gardens/kepaniw
www.gogobot.com › ... › Hawaii › Maui › Wailuku › Things to do
www.gohawaii.com/maui/...maui/iao-val...
www.gohawaii.com/**maui**/...
www.hawaiinaturecenter.org/
en.wikipedia.org/wiki/Iao_Valley
en.wikipedia.org/wiki/Iao_Valley/Wikipedia
en.wikipedia.org/wiki/Albizia_saman
en.wikipedia.org/wiki/Iao_Valley
en.wikipedia.org/wiki/Albizia_saman
good maui guide map: http://mauimapphotos.com/

www.ingramcontent.com/pod-product-compliance
Lightning Source LLC
Chambersburg PA
CBHW081457220526
45466CB00008B/2686